Winner of the Jules and Frances Landry Award for 2016

BAD GIRLS
at Samarcand

SEXUALITY AND
STERILIZATION IN A
SOUTHERN JUVENILE
REFORMATORY

Karin L. Zipf

LOUISIANA STATE UNIVERSITY PRESS
BATON ROUGE

Published with the assistance of the V. Ray Cardozier Fund

Published by Louisiana State University Press
Copyright © 2016 by Karin L. Zipf
All rights reserved
Manufactured in the United States of America

First printing
Designer: Michelle A. Neustrom
Typeface: Whitman
Printer and binder: Maple Press

Library of Congress Cataloging-in-Publication Data
are available at the Library of Congress.

ISBN 978-0-8071-6249-1 (cloth: alk. paper) —
ISBN 978-0-8071-6251-4 (pdf) — ISBN 978-0-8071-6250-7 (epub) —
ISBN 978-0-8071-6252-1 (mobi)

The paper in this book meets the guidelines for permanence and
durability of the Committee on Production Guidelines for Book
Longevity of the Council on Library Resources. ∞

CONTENTS

Illustrations follow page 62 and 106

ACKNOWLEDGMENTS

ON A HOT JULY DAY IN 2008, my friend Leslie R. Miller accompanied me on a road trip to Samarcand Manor, which at that time still operated as a youth detention center for girls. "Do you see any sign of it?" I asked her as I navigated the car down rural Samarcand Manor Road. We had driven for two miles, and still no reformatory in sight. There was an abundance of trees—oak and pine. An occasional unmarked dirt road intersected our route. I wondered aloud how a girl might escape from this juvenile delinquency institution in the middle of nowhere. "I don't know," voiced Leslie, contemplating. My friend, who had worked in the prison system in Georgia, was my only enthusiastic volunteer for this trip. While my other university colleagues jetted for research in exotic and sometimes dangerous locales, such as Paris, Berlin, Rome, Cairo, and Jerusalem, Leslie and I headed for Eagle Springs, North Carolina.

Eagle Springs is located in the heart of the North Carolina Sandhills. The landscape, sandy and hilly, embodies the name. The town lies one hundred miles south of Durham and twenty miles from Pinehurst, home of the luxury golf resort. Yet Eagle Springs does not bear the international cachet of its famed neighbor. The town is largely rural and agricultural, perhaps not much different than it was eighty years ago. There were no golf courses. The only signs of modernity had appeared a mile ago, where the road intersected the highway. There were railroad tracks, a sign for a bed and breakfast inn, and a melon stand, none of which are hallmarks, exactly, of the twenty-first century.

We passed a tobacco field, and I knew we had entered the past. "Is that broccoli?" Leslie asked. I slowed the car and pulled into a road that divided the field. Tobacco rose to our chests. In July, workers "topped-off" the blossoms and removed the leaves from below over a period of weeks until only the stalk remained. I imagined that in the 1920s and 1930s this entire two-mile stretch was covered by tobacco. In these fields a girl runaway might hide, but not for long. Finally we arrived at Samarcand Manor. Nearly all of the buildings we encountered at Samarcand were built in the 1950s and 1960s. Almost nothing was left of the original buildings, except the lovely little chapel built in 1926. Most had been destroyed by fire.

I would not have been able to write this book without Leslie. She generously lent me her skills and knowledge at this most unusual archive. Formerly a teacher in the Georgia prison system, her insight helped me understand how to navigate the institution, communicate with a staff actively managing a population of juvenile inmates, and step aside at the first sign of trouble. As a historian and archivist, her research and analytical skills led us to follow one historical question after another as we performed our research in the facility's small library. Leslie Miller is a fine scholar, humanist, and friend.

A variety of institutions provided funding for research and access to materials so that I could write this book. East Carolina University supported travel and offered a reduced teaching load. The ECU Faculty Senate provided significant support in the form of a Research and Creative Activity grant. Thanks to the Thomas J. Harriot College of Arts and Sciences and my colleagues in the Department of History, who recognized the project by awarding me the Lawrence F. Brewster Fellowship for a semester research leave. Staff at the North Carolina Collection and the Special Collections Department at the J. Y. Joyner Library offered their archival skills to locate important documents, allowed my use of digital and photocopy equipment, and helped procure other materials through interlibrary loan. Staff at the Laupus Library at the East Carolina University Brody School of Medicine supported two lectures on the project through the school's Medical History Interest Group Lecture Series. The North Carolina State Archives offered critical archival expertise in the State Library, the Office of Digital and Photographic Services, and in the Search Room. I am thankful for the rigorous expertise of the editorial staff and readers at the *North Carolina Historical Review,* which published an earlier version of chapter 2. The Southern Historical Collection at the University of North Carolina at Chapel Hill and the Special Collections Department at Duke University offered valuable assistance in locating and accessing manuscript collections. I reserve deep thanks to the staff, including Don Burns, at Samarcand Manor, the North Carolina State Home and Industrial School for Girls, who provided office use and access to the reformatory's library. Also I am grateful for the readers and editorial staff, including Rand Dotson, at the Louisiana State University Press. Their editorial expertise, and the indefatigable work of my copy editor, the keen-eyed Derik Shelor of Shelor and Son Publishing, transformed my manuscript into a beautiful book.

Special thanks go to colleagues and friends who read the manuscript and engaged me in discussions about my work. Their enthusiasm about the project helped to inspire the storytelling in much of the manuscript. Cheryl Dudasik-Wiggs, Susan Pearce, Donna Kain, Marieke Van Willigen, Holly Mathews, Anna Krome-Lukens, Angela Thompson, Anoush Terjanian, Mona Russell, Kennetta Perry, Lynn Harris, Christine Avenarius, Laura Mazow, Megan Perry, Jamie Leibowitz, and Keiko Sekino challenged me to study gender and science in the context of juvenile delinquency. Melissa Nasea, Daniel Goldberg, and Chandra Speight offered important medical perspective and friendly encouragement. The ECU Women's Studies Program hosted me for several presentations in its "Gender to a Tea Series." Avi Sickel used his detective skills to look for any living signs of the convicted arsonists. He found none. Vera Tabakova, Alethia Cook, Diane DeGroot, Olga Smirnova, Rose McMahon, and Melissa Tilley provided insight and supportive friendship. Paul D. Escott and Peter Charles Hoffer both offered valuable support and constructive critique for publication of the manuscript. I am deeply grateful for their professional advice and mentorship over the years. Alan White, Gerry Prokopowicz, Don Parkerson, Brad Rodgers, David Dennard, and Chuck Calhoun lent their encouragement in the form of letters of support and general publishing advice. Melton McLaurin, Elizabeth Leland, Ansley Wegener, Doug Brown, Kim Andersen, Ruth Cody, Chris Meekins, Anne Miller, Earl Ijames, Katrina Person, Ingrid Meyer, Rebecca Futrell, and Gabriela Baluskova Knox all contributed to this project. I am indebted to the work of some very bright graduate students, including Allison Miller, Virginia Dodd, Matt Crain, Matt Esterline, Scott Duryea, Heather Marie White, John Jarrell, Morgan Shaneil Pierce, and Adam Stoddard.

Good friends knew when to encourage the project and also when to leave it alone. Gisella Verando and Jennifer Haas helped inspire ideas early in the project. More thanks to my fine bandmates, who endured my crazy schedule but continued to play music with me: Jim Watson, Bob Aiken, Kevin Mills, Chris Jackson, and David Hayes. Sue Luddeke, my colleague, bandmate, and good friend, contributed a beautiful portrait for the book. She is a remarkable artist and a fine songwriter. The multitalented Robert Travas Hunter, read my work, prepared conference presentations, guest lectured in classes, and tutored my kids. Sue and Travas both are inspired artists and invaluable friends. Their creative contributions truly have enriched this book.

I extend the most sincere thanks to my family. Thank goodness my parents, Bob and Nancy Zipf, and my husband's parents, Louis and Lulu Sarris, warmly supported the idea of a book about the history of girl delinquents. I welcomed the incisive legal perspective of my sister and brother-in-law, Kristy and Kevin Green. My sons, Robert Ellis Sarris and Theodore Jon Sarris, remind me how blessed I am to be the mother of really great kids. Parenthood has had an important influence on my research. In my experience there is a relationship between lessons learned in family life and questions asked in professional historical research. My greatest appreciation I reserve to my husband, Jonathan Sarris. As a historian, he understands the "publish or perish" challenges of academia. He is a first-class historian, a great husband, and also an incredible father. I am able to write and research thanks to his many sacrifices. Thank you, Jonathan.

BAD GIRLS
AT SAMARCAND

Introduction

THE PAPER DRESS USED FOR SEWING assignments hung in a bedroom closet at Chamberlain Hall. Chamberlain was the "bad girls" dorm at Samarcand Manor, a 240-acre juvenile detention center for white girls in Eagle Springs, North Carolina. Margaret Pridgen, a teenage inmate, opened the closet door and lit a match. Then she lit the paper dress afire. It seemed the thing to do, to set a blaze that might force a girl's release from the reformatory. The flames quickly crackled the paper dress and turned it sooty black. Margaret shut the closet door and stood there to see that the fire continued to burn. Then she ran from the room and exited the building. At least five other girls remained in the building, locked in their rooms as punishment for their various teenage infractions. When Margaret Pridgen struck the match, she had no idea that in North Carolina arson was a capital offense. Her father had left her at Samarcand two years earlier, records said, for "running around with boys." She had merely hoped that she could, she said, "get out and go home."[1]

Pridgen was not the first girl to set a fire that night. Other girls, frustrated by the departure of some favorite teachers and in protest against disciplinary procedures that included brutal whippings and severe conditions in isolation, had fired the dorms, too. Hours earlier, Marian Mercer and Margaret Abernethy had set a stocking on fire in the attic. They had hoped that the blaze might force their release from the reformatory. Staff soon discovered this first fire and extinguished the flames. She did it, Abernethy said, because "I thought that if I set it on fire they would send me out, and I was tired of that place." Abernethy had served twenty-nine months not for any criminal offense, but because her father had sexually abused her and the court had nowhere else to put her. Other girls took the blame in hopes that their confessions would have them expelled from the institution. Virginia Hayes, another young defendant, said later that she had confessed because she would rather be in jail than at Samarcand.[2]

Pridgen's effort met with success. On March 12, 1931, the fire she set completely destroyed two residence halls at the State Home and Industrial School for Girls, popularly known as Samarcand Manor, located in the Sandhills of North Carolina about one hundred miles south of Raleigh. All escaped, but the fire risked lives and caused damage estimated in 1931 newspaper accounts of between $100,000 and $200,000. In the early 1930s, everything seemed aflame as North Carolina and the rest of the nation reeled from economic depression. Tension was high as unemployment soared, violence erupted amid employee strikes at local cotton textile mills, and the newspapers obsessed with reporting suicides of men and women caught in the depression's despair. The county sheriff immediately arrested fifteen girls on charges of arson. The next day he returned to the reformatory and arrested another, for a total of sixteen girls in police custody. They were incarcerated for the crime of arson on a habitation, which merited the death penalty in North Carolina.[3]

While awaiting trial, the girls committed arson twice more in county jails. Reformer and attorney Nell Battle Lewis took up their defense, a task made all the more difficult because the North Carolina General Assembly stripped them of their juvenile status to allow the governor to incarcerate them at his discretion. After controversial reporting in the press, the girls stood trial in May 1931. Nell Battle Lewis pleaded for a reduced sentence by arguing institutional negligence and a variation on the modern version of an insanity defense, claiming that her clients were "mental defectives" and clinically feebleminded. The judge discharged two girls and suspended the sentences of two others, then dismissed the accusations of neglect, reduced the sentences of the remaining girls, and called for an investigation. He committed twelve girls for a maximum five years in the state's Central Prison, where they were relegated to the cells above death row, the only fireproof wing in the facility.[4]

The early history of Samarcand Manor and accounts of the 1931 fire and trial provide a wealth of evidence that reveals an important turning point in North Carolina's public policy history. This turning point, dated 1931–33 and culturally represented by the triumph of eugenic science, marks a shift in cultural perceptions about southern white womanhood that had vast implications in public policy related to juvenile justice, incarceration, and sterilization. *Bad Girls at Samarcand* argues that the 1931 trial and resulting investigation at Samarcand created an opportunity for eugenicists and mental hygienists in state government to jettison Victorian models of reform and to formulate new eugenics-based classification policies for juvenile reformatories that reflected a class-based white purity campaign focused specifically on controlling the sexuality of poor women and girls.

Victorian-era reform policies predominated in turn-of-the-century North Carolina. Early reformers once characterized wayward girls as fallen victims who might be redeemed. In 1917, North Carolina reformers successfully lobbied state legislators to create a juvenile delinquency institution to serve white girls. No such state institution existed for African American girls. Before the Great Depression, many North Carolina authorities and reformers advocated a white supremacy that rested on the belief that white women, across class lines, possessed innate characteristics of piety, chastity, and gentility—virtues that elevated them over the women of any other race. Reformers founded Samarcand Manor to restore ladyhood to poor white girls.[5]

In the 1920s, some reformers began to refer to themselves as "mental hygienists." They viewed delinquent white girls as an unredeemable lower caste of feebleminded poor that required testing, institutionalization, and sterilization. Mental hygienists, a branch of social workers that embraced eugenic ideas, had begun a steady creep into state government. Eugenics is a set of beliefs that advocates improvements in human genetic traits by controlling reproduction among those considered "unfit." These social reformers promoted a scientific approach to public policy based on psychological classification of the poor and surgical sterilization of white girls and women. New state policies, including sterilization, emerged that separated affluent whites from the lower-class white girls they had once embraced. Sterilization already raged nationwide. Michigan first began forcible sterilization in 1897. North Carolina followed suit in 1919, but did not pursue actual sterilizations for another decade. Victorian-era policies fully gave way to a sterilization program with teeth in 1933. Women who refused to conform to the rigid characteristics of white womanhood suffered derision as feebleminded, incarceration as prostitutes, and sterilization by the State Eugenics Board. That year, the Eugenics Board instituted a routine classification system that screened and subjected Samarcand girls to sterilization. No longer did authorities see the Samarcand girls as prospective ladies.[6]

Girls constantly challenged these categories, which in turn influenced the behavior of the authorities. Girls often resisted convention by claiming their own sexual autonomy or just a modicum of privacy. Some ran away, set buildings on fire, or resorted to insolence. Adolescent girls, especially those of low-income families, rejected the Victorian model and refused to conform to Victorian standards of womanhood. When the Samarcand teenagers ran away or fired their bunks, they did so "to get out of this place"—their condemnation of the reformatory system. Their trial and subsequent behavior in jail prompted a crisis in the conversation about beliefs about white women, girls, and juvenile reform. When the girls refused the role of "fallen" victim, reformers and state authorities found it easier to abandon the Victorian standard, too. The girls' resistance against Victorian norms challenged reformers, some of whom responded with eugenicist language that charged the girls as "mental defectives." It was to the girls' misfortune, however, as by 1933 those in power embraced eugenic science and sterilization. Even so, girls adapted. They understood the power of the rhetoric about sexuality and sometimes employed the language to suit their own ends. This study will show that girls fully participated in reinforcing homophobic fears about sex-segregated correctional institutions. For example, some girls resisted reformatory commitment and eventual eugenic sterilization by invoking the language of homophobia with judges, thus negotiating a shorter term in prison instead of an indefinite term as required in the reformatory.[7]

This book examines attitudes about juvenile delinquency and correctional institution policies, particularly those related to adolescent white girls before, during, and after the trial. It argues that the Samarcand arson trial and subsequent investigations (1931–33) marked a turning point in the adoption of eugenics policies in North Carolina's emerging juvenile court program and reformatory system. These new policies, specifically classification, sterilization, and parole, were intended for the purposes of controlling white girls' reproduction in ways that had not existed before the trial. By specific study of Samarcand Manor, the book examines the complicated relationships and alliances among reformers to establish the institution. It carefully fleshes out the eugenics ideology of influential women associated with the institution, including Samarcand Manor founder Kate Burr Johnson (later the first female public welfare commissioner), Samarcand Manor superintendent Grace M. Robson, and the girls' defense attorney, Nell Battle Lewis. Lewis, a high-profile attorney, developed a legal defense that included testimony to the court that her clients were feebleminded, a move that contributed to the changing public perception of the girls who lived at the reformatory.[8]

Although the book focuses on juvenile delinquency, state-sponsored sterilization is a critical part of the story. Eugenic sterilization programs began in earnest at the Carnegie Institution's Cold Spring Harbor in Long Island during the 1890s, spread across the United States through corporate organizations such as the Rockefeller Foundation, found support in local and state organizations and politics, and eventually made their way to Germany. The U.S. South came late to the program; North Carolina passed its first meaningful sterilization law in 1929. The impact was global. The United States served as a precedent for the German sterilization program. German and U.S. eugenicists shared a common ideology and drew upon one another for support. Adolf Hitler catapulted eugenics into the platform of the National Socialist Party, and there it outpaced the United States. One Virginia eugenicist remarked in 1934 that "the Germans are beating us at our own game." In 1963, the North Carolina sterilization program rose to national prominence when John F. Kennedy appointed Ellen Winston, North Carolina's commissioner of public welfare, as the first U.S. commissioner of public welfare. By the 1970s, North Carolina ranked third in number of victims among thirty-three states that sterilized over sixty thousand people. Victims included the urban and rural poor, African Americans, Jews, Mexicans, criminals, alcoholics, epileptics, and anyone seen as mentally ill from New England to Florida and from the Carolinas to California.[9]

A story about juvenile delinquents and sterilization policy is tricky to research; it requires access to sources that institutional authorities are not inclined to share. This cultural history is steeped in legal sources, especially documents related to the juvenile courts, juvenile justice, and delinquency. Legal developments are critical to explaining how this story unfolds. Any history of adolescence is difficult to trace because state laws prohibit researchers from accessing juvenile records—especially those related to individual juveniles, such as the Samarcand defendants. Therefore, this study required creative use of state records, including those maintained by the State Board of Health, State Board of Charities and Public Welfare, State Charitable, Penal, and Correctional Institutions, Department of Juvenile Justice and Delinquency Program, the State Highway and Public Works Commission, and the Governor's Papers. By close study of these records and verification through newspaper and manuscript collections that date from the 1920s to the 1940s, this book reinforces Johanna Schoen's study of the state health and welfare programs and extends it to say that North Carolina's juvenile court system was steeped in the cultural assumptions of eugenics and Jim Crow, ideologies that defined class, race, and sex differences in that era.

By 1933, Samarcand no longer served as a training school for southern ladies, but as an asylum for girls of low-income backgrounds whose wayward behavior elite southerners interpreted as feeblemindedness. The state had outlawed whippings, but had institutionalized a more nefarious cruelty: sterilization. Through the tale of their crime and trial, this book narrates the cultural and legal strategies that bad girls used to navigate the precariousness of life in the age of eugenics.

1

A Place for White Girls

THE TRICKY POLITICS OF JUVENILE REFORM

IT WAS MARCH 7, 1917, AND in the gallery of the General Assembly sat Birdie Dunn, a registered nurse. She was there to speak her mind against a bill to create a reformatory for white girls. The bill's author, Mrs. G. Hope Summerell Chamberlain, president of the North Carolina Federation of Women's Clubs, also attended the session. Chamberlain's goliath Federation of Women's Clubs had labored tirelessly for the bill and a $25,000 bond measure earlier that winter. In fact, Chamberlain had written the legislation herself. Representative Walter "Pete" Murphy, the House Speaker and Chamberlain's childhood friend, graciously agreed to present the bill to the legislature. His pleas to save the "fallen" women of North Carolina were met by resounding applause from the legislators on the floor and women in the gallery. Hoping to offset some of Mrs. Chamberlain's momentum, Dunn had recruited her own representative from Anson County to read her letter of opposition. As the applause subsided, R. E. Little rose to the podium. Then he delivered Dunn's minority view. "I have been a nurse for 20 [*sic*] years, and I have looked upon all places of human misery, and this seems to me a most one-sided charity, totally impracticable and offering no rehabilitation of individual or family." Where Chamberlain viewed girls on the street as victims of predatory men who had failed their role as protectors, Dunn viewed wayward girls as unfortunate victims of their social environment.[1]

In 1917, no consensus existed on how to handle juvenile delinquent girls. Multiple voices proposed solutions that at heart contested the meaning of southern white womanhood, an ideal that not only represented beliefs regarding gender differences, but also implied views about civilization, race, and women's relationship to the state. In the early twentieth-century United States, a battle raged between evangelical reformers who preached the Social Gospel and a professional class of social workers who objected to the moralistic evangelicalism of the female reformers. Reformers debated how to curb white girls' sexual behavior. Evangelicals viewed these white girls as unfortunates who had "fallen" and required uplifting. Social workers saw them as wayward girls led astray by a bad environment and requiring state intervention. The 1917 debates in the North Carolina General Assembly demonstrate that transitional figures—that is, professional social workers with women's club experience— bridged the gap between evangelicals and professionals by advocating a white girls' home that would rescue "fallen" girls from their environments, educate them, and restore them to southern ladyhood.

The meaning of white womanhood strongly influenced the protracted conflict over adolescent female delinquents in the transfer of power from evangelical reformers to social workers. This chapter examines contested expectations of white womanhood that played out in the conflict between reformers and professional social workers in North Carolina. The conflict centered upon the problem of what to do with white girls—that is, runaways and delinquents— who roamed the streets. Race was central, as reformers only discussed the plight of girls identified as white. White supremacy rested upon the idea that white men possessed sexual exclusivity to white women. Street or "problem" girls threatened this exclusivity by their unrestricted access to all sorts of partners.[2]

The plan for a state reformatory for "fallen" white girls originated with a man, not a woman. A minister had conceived the idea three years earlier. The Reverend A. A. McGeachy of the Second Presbyterian Church in Charlotte, North Carolina, breathed the Social Gospel. But he was no zealot. He represented a significant religious wave that had moved across the nation in the early twentieth century. Social Gospelers professed that industrialization had created a paradox of good and evil. While industry and machinery created wealth, knowledge, and efficiency, the unchecked capitalism that accompanied this progress also caused physical suffering, poverty, and injustice. Industrialization thus had created an evil: a society divided into two classes, rich and poor. According to the tenets of the Social Gospel, it was the duty of good Christians to reach out to the poor by helping to alleviate the evil consequences of industrialization.[3]

McGeachy laced his sermons with the creed. His plan was to raise awareness in his congregants of social reform. Rather than focus on individual salvation, he said, some Christians had begun to stress humanitarian work in health and sanitation, lobbying for the living wage and work for the unemployed. In McGeachy's own words, these humanitarians "feel that perhaps a bath every day is not less important than a baptism once in a lifetime, that a clean body bears some effective relationship to a clean conscience." These Christians, McGeachy explained, had begun to think seriously about the Epistle of James and the doctrine of "good works." While still loathe to replace "faith" with "works," social reform Christians found value in following Jesus's example in their search for His approval. To illustrate this point, McGeachy explained the story of the Samaritans in Luke 9:53. When Jesus came to the village of the Samaritans, they spurned him because they did not like his face, as it was "turned to Jerusalem." McGeachy reproached the Samaritans for such unchristian-like behavior. A good Samaritan, he argued, would embrace even the most anguished faces in His kingdom. Jesus's approval would follow. Then His face, McGeachy professed, "the face that was turned toward Jerusalem, the face that once, wet with tears, bowed low over a weeping Magdalene, would then be turned to us."[4]

In March 1914, McGeachy preached to his flock on the white Mary Magdalenes of North Carolina. He implored his audience to find empathy for "fallen" women, seduced by the streets into lives led in sin. "In North Carolina," he professed, "we have no reformatory for fallen women." With restricted exceptions "there is no place where a woman who has lost her womanhood can find it again." Even the most penitent woman, he argued, had little chance for redemption. "We have a place for criminals—the penitentiary; a place for the insane—the asylum; a place for the indigent—the poor house; a place for orphans—our orphan homes; a place for the sick—our hospitals; a place for the blind, the deaf, even the feeble minded. But no place in the State, with all its charities and philanthropies, for a woman who has sinned and is sorry, to find a helping hand."[5] Sin, sadness, and sorrow—this was the triad that McGeachy used to describe the victimization of women. McGeachy drew heavily upon a common language of womanhood adopted by evangelical female reformers from the previous century. Nineteenth-century reformers had identified a "cult of true womanhood" that characterized women as naturally pure, pious, domestic, and submissive. Women's passivity at once made them desirable to men but also made them vulnerable to the most sexually aggressive and unscrupulous of the opposite sex. Popular literature commonly referenced the melodramatic story of women's seduction and abandonment. In popular seduction narratives, such as *Charlotte Temple* (1791) and *Maggie, Girl of the Streets* (1893), the female protagonist typically falls in love with a sexually irresponsible man who impregnates her, then abandons her to a life of prostitution. Eventually she dies of sorrow, shame, or suicide. By using this sexualized narrative, evangelical reformers created a new language of women's sexuality that allowed them to reclaim the respectability of female sex workers and redeem them.[6]

McGeachy relished the melodramatic narrative. He employed it in several anecdotes for his March 15 sermon. In each story, the sins of the woman are nebulous, but her plight afterward is terrifically horrifying. In one, a young girl killed herself by eating broken glass. He references her transgressions only vaguely by asking his audience to "imagine the state of a poor woman's mind when she hails a broken bottle as a friend." Yet, according to the good reverend, she experienced a most unholy demise. As she lay dying in a disreputable part of Charlotte, unscrupulous men gawked voyeuristically: "The doors and windows of the room where the girl lay dying were open all night. Men looked through those windows, men tramped through the room with their hats on—why should they remove them? Men sat watching with dull curiosity the contortions of her body, the writhings of her face, and there was no decent place in which she could die, nor quiet hour in which she could prepare to meet her God." Another anecdote describes the funeral service of a girl who killed herself with carbolic acid. A third story tells of an eighteen-year-old girl who was abandoned by a man in a Charlotte hotel and forced to find refuge in the red-light district "not because she was criminal, but unfortunate; not because she wanted to go there, but had no other place to go." All of these stories carried the same theme. Women's transgressions went unsaid, but the social consequences of their actions were excessive and unfair. In one anecdote, Reverend McGeachy described his visit to the Charlotte police station, where a woman, crouched in a cell, awaited the birth of her child. "There where drunkards had vomited," he breathlessly explained, "and beat their bloody hands against the bars in delirium, she awaited the most sacred experience of womanhood, and there was no place, not even the jail, to which she could be legally committed."[7]

Thus, reformers like McGeachy viewed "fallen" women—that is, prostitutes and unmarried mothers—as candidates who could be reformed to their "natural" state of womanhood. McGeachy pleaded with his congregants to launch a movement for "these sad-faced sisters of ours." He envisioned a state institution "big enough for all and beautiful enough to attract the penitent," where unfortunate girls and women could seek the restoration of their natural state of womanhood. Girls would live and learn at the institution. Reformers would emphasize domestic skills or wage-earning skills befitting their sex. "Suppose we teach them there domestic accomplishments that will fit them to preside over homes of their own—if they ever have homes—or employments that will enable them to earn an honest livelihood."[8]

McGeachy's message and melodrama deeply touched one experienced reform advocate, Hope Summerell Chamberlain. A friend had sent her a copy of McGeachy's sermon, and she immediately enlisted him as her ally. She was powerful because she served as the voice of a network of women's clubs that had lobbying strength not only at the local and state level, but at the national level as well. Women's club activity had originated in the nineteenth century with the Woman's Christian Temperance Union and with missionary, reform, and charitable societies that enabled women to express a new ideology of benevolence. According to this belief, women shared unique characteristics of morality, virtue, and benevolence that they could use to help shape the new political and social structures of the nineteenth century. The rhetoric of benevolence applied to women as a group and sometimes obscured class and race identities. In twentieth-century North Carolina, middle-class black and white women sometimes transcended the color line to promote interracial cooperation on causes such as temperance and women's suffrage.[9]

In North Carolina, this "awakening of womanhood," as one reformer described it, was a revolution, "bloodless but not purposeless." According to Sallie Southall Cotten, a contemporary of Chamberlain's, women were steadily gaining education, knowledge, and experience and had begun to heed the call to contribute to the "onward march of civilization." "Educated womanhood," Cotten described in the *History of the North Carolina Federation of Women's Clubs, 1901–1925*, was "absolutely essential to develop 'the female of the species' into proper mates for educated manhood, and for the fulfillment through them of God's law of evolution which forever calls for higher types." Organized womanhood, Cotten explained, enabled woman to break from the most "restrictive conventions of her isolated individuality" and apply her feminine virtues to the progress of civilization. The club movement marked the awakening of womanhood to her new feminine role. By 1914, North Carolina women had founded numerous reform efforts, including public schoolhouse reform, health and sanitation reform, and suffrage associations. Many of the clubs espoused evangelical values, but others offered a protofeminism that encouraged women's activity in public spaces within their conventional roles as mothers and homemakers.[10]

Chamberlain was an unreconstructed southerner steeped in Lost Cause ideology. Highly educated and possessing uncommon self-reflection and introspection, Chamberlain matured into a prolific writer, one who espoused white supremacy. Among the many writings that she produced was her own autobiography in 1938, entitled *This Was Home*. Much of the book is a history of her family and community in Salisbury, North Carolina. She describes her own birthdate in 1870 as a milestone in Reconstruction history. She was born, she said, on the very day that the "renegade" Republican governor W. W. Holden had ordered a mountain militia known as "Kirk's Army" to march against the Ku Klux Klan. Chamberlain described Kirk's Army as "a body of Union guerrillas recently disbanded, recruited with rough mountain boys spoiling for a fracas." The Ku Klux Klan, for its part, was charged with several murders "not at the time, then or since, proved against it." The altercation never occurred, she alleged, because the pro-Union militia had fallen upon a stock of whiskey. Southerners like she would have to wait impatiently for the "dreary comedy of the reconstruction government" to end. Elsewhere, she describes the mystical rise of the Ku Klux Klan by young men who came home and "witnessed the fantastic tricks of the Freedmen's Bureau, and saw how the gaping Negroes were by it indoctrinated in hatred to their former masters." It almost goes without saying that even as late as 1938 her politics idealized the Confederacy and the Lost Cause.[11]

This white supremacist imagery lay at the core of southern whites' desire for a racially segregated girls' institution. White womanhood rested at the heart of Chamberlain's racial hierarchy. The conventional wisdom of the day characterized white women as virtuous, pure, genteel, and the exclusive object of white men's affections and desires. Black women, however, could not be ladies. Whites viewed them as naturally lustful and licentious. Southern sex and race relations had produced a multidimensional double standard where white women represented gentility and purity and black women symbolized sexuality. White men could engage in sexual liberties across racial lines and black men could not. "The South's multidimensional double standard," one scholar has written, "elevated white men, degraded black men, and treated all women as objects—those on one side of the color line to be cherished, those on the other to be used." To maintain this fiction, southern whites needed to elevate the lowest class of white girls to the status of ladyhood, if only in name. Neither Chamberlain nor other white women reformers offered to sponsor an institution for African American girls.[12]

Yet, for all that, Chamberlain claimed to be a modern woman. Her views on gender, she wrote, grew from the "feminist heresy" of her grandparents. They had, she argued, "determined that their three girls should have instruction in any branch which would have been open to them if they had been boys." They educated Chamberlain's mother and sisters in botany, geology, Greek, Latin, and history. Her family rejected the "die-away, romantic ideal for women" and encouraged athleticism and constant exercise. Chamberlain wrote that she believed that women were equal to men in "mentality, but they are thousands of generations behind them in training." The suffrage movement grew, she observed, as women began to question the old authoritative rules. Women had excelled with "all the virtues that men considered irksome," and had silently begun to doubt "if the sauce of the gander and the goose were so different" when the flesh was just the same. As women asserted themselves at home and in the world, she argued, "[men] will begin to prize us as partners and equals."[13]

Unlike some of the most outspoken feminists of her day, Chamberlain limited her feminist critique to her observations on spousal relationships. She did not go so far as to challenge marriage laws or the system of coverture that subordinated wives to their husbands. Rather, she limited her remarks to observations on the inequalities that existed within marriage. She criticized husbands for taking no interest in their wives' pursuits. She also noted the patronizing nature of the spousal relationship that required a husband to protect and provide for his wife, but prohibited her from earning by "real work one single penny which she can call her own." Women resented it, she said. The men, she observed, "have no idea of the seething inside many a quiet exterior." As for her own relationship, she prized her husband's respect for women. Her husband was from New York. When they first met, he had impressed her with his beliefs on gender equality. Apparently he warmed her heart with his respect for women. "He did not call us either 'females' or womanish," she fondly recalled, "but said plainly that women's minds were actually equal to men's so that they learned quite as easily and made better grades." His charms struck her as gentlemanly, and she agreed to marry him. "Yankees did not always have horns and tails," she said of her northern-born husband.[14]

Though an armchair feminist, she served as a foot soldier in the North Carolina Federation of Women's Clubs. Chamberlain made 1917 her year and social reform her cause. She assumed an especially active role in the North Carolina Federation of Women's Clubs. That year her peers elected her chairman of the Legislative Commission, where she pushed not only for a white girls' juvenile delinquency institution, but also lobbied for an improved and stronger State Welfare Commission. To Chamberlain, it appeared that the 1917 North Carolina legislature was composed of young, modern men with reformist principles like her own. Forward-looking, progressive, and constructive, these men, she believed, would help her carry through her agenda. She also solicited help and advice from other like-minded women, including social worker and fellow clubwoman Kate Burr Johnson, who was elected president of the North Carolina Federation of Women's Clubs that same year. Another important consultant on the project was a prominent social worker from Philadelphia named Martha P. Falconer.[15]

Chamberlain probably related well with Kate Burr Johnson. Johnson had served many years in charity work as president of the Raleigh Women's Club and later in the North Carolina Federation of Women's Clubs. The ideology of benevolence thus shaped her earliest experiences in social work. One commemoration of her tenure as president of the NCFWC describes her role as lifting social reform to a new plane. Clubwomen in North Carolina previously had focused on charity projects, such as schoolyard beautification and sanitation. As president, Johnson shifted the focus of women's civic reform from building flowerbeds to more politicized activities, such as selling and buying war bonds, legislative lobbying, and public welfare improvements. Johnson's leadership inaugurated a new role for women reformers. She had shifted her attention to social service because she believed social work to be the next stage of women's reform efforts. In 1919 Governor T. W. Bickett appointed her director of the Bureau of Child Welfare of the North Carolina State Board of Charities and Public Welfare. She thus served as a crucial transition in women's reform from benevolence to the professionalization of the field. As clubwomen who once shared the ideology and work of benevolence, Chamberlain and Johnson had much in common. In fact, Chamberlain greatly admired Johnson. Her correspondence to other prominent women reformers celebrated Johnson's 1919 nomination and appointment. In 1921, Bickett appointed Johnson public welfare commissioner. By the 1920s, professions such as social work offered a credo that heralded individual merit and education, thus attracting women to their ranks. Johnson's state appointment inaugurated a national calling of women, especially educated social reformers, into the social work profession.[16]

Yet Chamberlain might have paused at Johnson's quite provocative—indeed, titillating—views of girls and sexuality. Here, Johnson's experience as a social worker (in other words, her scientific approach to social problems) emerges. In 1926 she wrote an article in the *Journal of Social Hygiene* in which she hardly shied from the topic of sex. "The problem of sex has created a moral double standard unfavorable to girls. No honest, thoughtful person," Johnson claimed, "can deny that we have had an abnormal repressive attitude toward sex, or that we are now witnessing a time of license and revolt." Sex, Johnson argued, is not sinful but "a fact—a great, beautiful, absorbing, interesting fact." It serves to inspire, she said. "In its transmutation, its sublimation, it is the spring of many of our highest aspirations and the motive power of much of our most valuable creative activity." She advocated sex education. Parents and teachers could resolve girls' delinquency, she argued, by developing intelligent programs in sex education and recognizing that immorality was not a sin but a sickness of the feebleminded and the mentally defective. Chamberlain clearly appreciated Johnson for her experience in women's clubs, but it is not known how she received Johnson's position on girls, sexuality, and mental deficiency.[17]

Unlike Chamberlain and McGeachy, who evangelized white women's protection, Johnson emphasized the need to protect and segregate the feebleminded from "normal" members of society. Fallen women, she conceded, required protection. However, she also believed that many of these women also possessed undesirable traits of mental deficiency and feeblemindedness. As a reformer she encouraged the creation of institutions such as Samarcand Manor and Caswell Training School for the Feebleminded to segregate women with these presumed undesirable traits. Later, as commissioner of public welfare, she directly advocated and lobbied for the state's first eugenics laws, which established sterilization and the Eugenics Board. As a longtime member and former president of the North Carolina Federation of Women's Clubs, Johnson held close friendships and alliance with women such as Chamberlain across the state. Reform-minded women were an integral part of Johnson's social reform agenda. As these women lobbied for the protection of women, children, and the mentally afflicted, Johnson began to advocate stronger sterilization laws and more restrictive marriage laws.[18]

On the nature of southern white womanhood, Chamberlain and Johnson shared common ground but also expressed significant difference. Both women advocated the uplift of "fallen" white girls, and both agreed that through careful supervision and reform girls from the street could relearn their proper role as southern ladies. Both had presided over an alliance of women's clubs to initiate broad civic reforms. However, significant difference also marked the two women. Chamberlain's feminism did not challenge the role of the male protector, but argued only that he should become more sensitive to the needs and interests of his wife. Johnson, on the other hand, valued girls and women as agents in their own sexual expression. Sex was a natural state, thus promiscuity was not a sin. Rather, Johnson viewed promiscuity as a societal problem brought on by mental deficiencies that prevented a girl from controlling her natural impulses. Chamberlain sought to restore southern womanhood in fallen women who had succumbed to seduction. By contrast, Johnson sought to isolate and curb undesirable traits, such as promiscuity, that appeared in the most mentally defective white girls and to prevent the spread of those traits in the white race.

While Chamberlain admired Johnson, her relationship with Martha P. Falconer was probably a bit less comfortable. Falconer was not a clubwoman, but she possessed a far longer service in professional social work, and thus shared a different outlook than the other women on the nature of "problem" girls. She gained national prominence as the superintendent of the Pennsylvania State Industrial School for Girls, Sleighton Farm, and held a long association as a board member and advisor for the American Social Hygiene Association. As early as 1910, Falconer wrote on the distinct nature and causes of delinquency among girls. Work with girls was "more difficult" and "less hopeful" than work with delinquent boys, she explained. Girls, she argued, "are more emotional and less reasonable than boys." Troublesome and "often hysterical," they were difficult to help. Their delinquency, she explained, usually was rooted in social causes; delinquent girls came from broken homes, lacked positive sources of recreation, and lacked any formal training or schooling. Lack of proper care for the "feeble-minded girl" also contributed to the delinquency problem. Most of these problems, she acknowledged, beset boys, too. But society remained less forgiving of girls. "A boy may be grossly immoral and his immorality does not always find him out and follow him; it is not so easy for the girl who has been immoral to be helped back to a normal place in society." In fact, Falconer argued for longer sentences for delinquent girls than boys. She concluded that authorities should grant probation to girls with discretion "rather than to give her the freedom which she would abuse and perhaps be the cause of getting others into trouble."[19]

To Falconer, delinquent girls were a menace, especially in times of war. She was a social hygienist, a professional dedicated to wiping out vice and venereal disease. As early as 1914, she served as a director on the Board of Directors of the American Social Hygiene Association, an organization committed to the "educational, sanitary, or legislative" suppression of the "disease of vice." In 1918 she published her views on girls' delinquency in an article that neither Chamberlain nor Johnson could have missed, considering their legislative activity. Her byline listed her as "Director of the Section on Reformatories and Houses of Detention for Women and Girls for the National Commission on Training Camp Activities." Delinquent girls, Falconer proclaimed, were "a menace to the men in training" because the venereal disease they spread threatened the efficiency of the armed forces. Falconer advocated the idea of detention homes or clearing houses to hold all young women and girls except "hardened prostitutes and 'repeaters,'" to await trial and undergo medical treatment and physical and mental tests. Untrained, neurotic, irresponsible girls on the verge of drifting into prostitution required institutionalization in the countryside, where they could be put to work. Others, such as feebleminded girls and women, required permanent custodial care. Due process was not her cup of tea. Rather, she stressed the importance of institutionalization until adulthood or indeterminate sentences with the possibility of parole.[20]

Chamberlain probably recoiled at Falconer's use of the word "menace" to describe poor, troubled white girls. Some southern whites viewed mental hygiene and eugenic control as a challenge to white supremacy. Eugenic characterizations of poor whites as "white trash" challenged southern white supremacist views of a monolithic, superior white race. The Agrarians, an intellectual mélange of novelists, political economists, and political and social theorists that included William Faulkner, John Crowe Ransom, and Andrew Nelson Lytle, took their name in idealization of the southern agrarian traditions, the healthful effects of country living, and the virtue in yeoman farmers who lived off the land. Modernity, industrialization, and social fragmentation, not mental degeneracy, they asserted, had contributed to rural southerners' demise. Southern defenders of white supremacy such as the Agrarians bristled at mental hygienists' characterization of poor whites as being feebleminded or of "lesser stock." These stereotypes further eroded southern autonomy. Chamberlain, a clubwoman and suffragette, was not known as an Agrarian, typically a category reserved for men. However, the Agrarians vociferously invoked a variation of Lost Cause rhetoric in defense of the southern hierarchy that many elite southerners, male and female, would have found familiar.[21]

Falconer and Chamberlain also were dissonant on their views of race. Falconer shared no romanticized vision of white womanhood and rejected any melodramatic ideal of the young unmarried mother or prostitute. White girl sex offenders were hardly much different, her words implied, than African American girls. Both presented a sexual menace to society and to the men in the armed forces. Unlike white southern reformers, she advocated detention and reform for African American girls, and did not object to racial intermixing in these institutions. In fact, she implored states to step up and take responsibility for addressing the "menace of the immoral colored girls and women," a problem that directly affected white citizens "and is theirs to face as much as it is for the colored themselves."[22]

It is not hard to imagine that Chamberlain may have left some meetings frustrated and a bit bewildered. The three women allegedly worked well together, but clearly they did not share precisely the same outlook on girls, women, and social reform. While evangelical reformers and clubwomen viewed unmarried mothers and prostitutes as victims who were once seduced and abandoned, professional social workers adopted a more diagnostic and, one might say, a less humanized approach. Social workers characterized female sex offenders as "feebleminded" and "sex delinquent." They submitted their subjects to a battery of mental tests intended to measure their intellectual capacity. Then they ranked them by their degree of mental defectiveness. Chamberlain, the clubwoman, and Falconer, the professional social worker, epitomized these differences. However, Johnson shared common ground with both. Like Chamberlain, she envisioned unmarried mothers and prostitutes as victims of sexual conventions. But like Falconer, she advocated diagnostic study and classification of these girls that would have caused reformers like Chamberlain and McGeachy to pale.[23]

Yet that 1917 winter the women put aside their differences and crafted a far-reaching legislative agenda. They had a two-pronged plan. First, they would replace the 1905 State Board of Public Charities and Welfare law with new legislation that would create a more powerful State Board of Charities and Public Welfare. The old board was merely supervisory and wielded few powers. The new board, appointed by the governor and elected by the General Assembly, had the power not only to supervise but to inspect the whole system of charitable and penal institutions of the state. The legislature empowered the board to study a wide range of social conditions, including poverty and the treatment of prisoners. The new seven-member board also would be charged with "the welfare of dependent and delinquent children and to provide for the placing and supervision of this class." In fact, the board possessed the power of a legislative committee or court. It could subpoena witnesses, compel investigation, and demand documents. It was privy to the budgets and innermost workings of private institutions, such as orphanages, and could grant and withhold licenses of operation. The law specified that one seat on the board would be reserved for a woman and required the creation of a commissioner of public welfare to oversee a newly created Department of Charities and Public Welfare. In 1921, Governor T. W. Bickett appointed Kate Burr Johnson as the state's first public welfare commissioner.[24]

The second prong was their pet legislation, the State Home and Industrial School for Girls and Women. First, Chamberlain intended to keep the legislation simple in order to provide maximum discretion to a new five-member Board of Managers, "three of whom shall be women," to oversee the handling of the inmates at the reformatory. The Board of Managers possessed the power to purchase the real estate for the institution and to erect buildings. It would manage and control the superintending, maintenance, employment, and inmates in nearly all matters of discipline and management. These powers included the appointment of a superintendent and the classification, discipline, and power of parole and discharge of all inmates. The bill specified that any girl or woman arrested and charged as a "habitual drunkard, or being a prostitute, or of frequenting disorderly houses or houses of prostitution, or of vagrancy, or of any other misdemeanor," could be committed to the institution at the full discretion of the Board of Managers. Sentences would be indeterminate, but not longer than three years. Discharge was granted by the Board of Managers, not by a court. While the board could select its own inmates from county prisons, it would also accept applications by any white girl who "confessed herself guilty of any offense or any wayward conduct and may in writing express her desire" for reform at the institution. Other sections specified the board's power over runaways and the curriculum and management of an industrial school to provide each girl a "useful trade or profession and improving her mental and moral condition." It was no coincidence that in 1918 the governor promptly appointed Hope Summerell Chamberlain to this board. Chamberlain served until her health declined in 1920. Another bill included a $25,000 initial appropriation and $10,000 per year.[25]

Next came the promotional blitz. The North Carolina Federation of Women's Clubs did what it knew best. Women across the state turned out petition after petition in favor of the reformatory. Letters and petitions came from counties across the state. The Woman's Christian Temperance Union, the United Charities of Morganton, the YWCA, missionary societies, churches, civic groups, all sent petitions listing dozens of names each. In Elizabeth City, citizens gathered at a community service day signed a petition that called the bill "tardy justice" for girls and women. They turned their men out, too. Boards of aldermen, businesses, and ministerial unions sent petitions to the General Assembly. Two nights before the legislature met to debate the bill and appropriation, a flurry of telegrams arrived at the office of C. G. Wright, all from men in Charlotte who passionately supported the legislation. The first telegram arrived at 12:22 A.M. "We are counting on you," wrote J. B. Thompson. Eighteen minutes later, A. S. Mowrey followed with more words of support. At 2:25 A.M. Carl M. Turnfield wrote that it was about time North Carolina supported its own delinquent girls. The bill, he wrote, "is to correct and prevent delinquency it is not for hopeless cases. Vote for it we sent hundreds of girls to other states last year let us take care of our own." During the day, more telegrams rolled in. The last appeared at 11:15 P.M. when Jonathan R. Pharr wrote that "North Carolina ought to be ashamed to dump her unfortunates on other states." The women had created the momentum and men provided additional political force in favor of the legislation.[26]

All of this effort was in anticipation of the big debate in a joint committee of the House and Senate on February 6. Chamberlain had mustered her forces. She had a powerful senator, Alfred M. Scales, a prominent attorney from Greensboro, introduce both pieces of legislation. She also organized an impressive lineup of women presenters that consumed most of the afternoon. She spoke first, and in the language of the evangelical reformer. "This is an institution," she declared, "which would allow a girl who has gone wrong to be saved from her own ill doings." It was not a jail to house fallen women who would be released only to further their degraded lives. It was a reform school for the country girl led astray by urban ways. It was an institution for the town girl who found the streets more attractive than home. The goal, Chamberlain argued, "is to give these girls a chance . . . and to obliterate an evil almost as old as the world." Citing statistics from other states, she argued that a reformatory would save these girls. As part of her carefully orchestrated plan, she introduced a physician and the Charlotte chief of police to support her claims. Other speakers followed, including Edith Palmer Hutt (also known as Mrs. W. N. Hutt), who argued that the institution would help treat social diseases that would have long-term effects on girls' reproductive cycles. "One of the objects of the institution," she argued, "was to cleanse the social stream." Specifically, a white girls' reformatory would allow physicians to treat venereal diseases that caused "insanity and blindness to children born." Daisy Denson, secretary of the State Board of Charities and Public Welfare, presented "an armful of reports" to show that girls could be reclaimed. And Reverend A. A. McGeachy offered statistics to argue that in a study of thirteen similar institutions, "95 per cent of the girls . . . had been restored to good and useful lives."[27]

The proponents thought they had offered an ironclad case. But several senators objected to the provision that allowed any girl or woman arrested on any misdemeanor charges to be sent to the institution. Some senators wondered if counties would send prostitutes to the reformatory. "Yes," McGeachy replied, "and they would be reclaimed of their sins." Another senator noted the problem of housing misdemeanants and felons in the same facility. Virtuous girls arrested for minor infractions, he said, would be forced to mix with the degraded and fallen. That would not happen, retorted C. W. Tillet of Charlotte. "No judge would send any woman to the institution for violating the speed limit," he said. But the language of the law allowed for it, argued a senator from Buncombe County. Another legislator dismissed the matter as irrelevant. "Remember the story about the fellow in the distance who could see the fly on the barn door but could not see the door?" he inquired. The debate on misdemeanants was a mere technicality, he argued, that should not prevent the passage of the bill.[28] This line of argument proved prescient. Later, most girls committed to Samarcand were convicted of misdemeanors. Some inmates had no record at all.

Commitment to a reformatory, others argued, violated a girl's right to liberty and due process. While some senators objected to the issue of mixing hard prostitutes with young girls, others were most inflamed by the power that rested in the Board of Managers. According to critics, the board possessed an unlimited power that violated individual liberties. It had the power to determine length of sentences, arrest runaways, and reject girls from admission by county courts. Kate Burr Johnson stood to defend the indeterminate sentence. A Senator Jones strenuously objected. "I am jealous of the rights of a citizen and I am a suffragette of the rankest sort," he said, "when a person is convicted of a crime hasn't she the right to know the length of her sentence?" Johnson responded that the indeterminate sentence worked; it was the cornerstone at other institutions. She used the analogy of a hospital stay to reinforce her point. No physician would send a patient to a hospital for a definite length of time. The patient would remain until well. It was the same with these girls. Their length of stay must be left to the wise discretion of the management.[29]

Other legislators wanted answers to questions about jurisdictional powers over escapees. "What about runaways?" Senator Jones asked. The Samarcand Board of Managers, he said, should not have power to arrest a girl who escaped or broke her parole. An arrest warrant must be issued by the county magistrate, he argued. In the case of this reformatory, Jones complained, the management's police power was too far reaching. He reiterated his objections to the broadly defined powers of the proposed reformatory's Board of Managers in terms of jurisdiction, due process, indeterminate sentences, and police power. "Was it right for the Board of Managers to reject any woman sent by a judge?" he asked. "Was it proper to deprive a woman of her liberty?" The Reverend McGeachy answered with the most paradoxical response, "while you are right in every particular you are wrong all through." For the evangelical reformer like McGeachy, priority rested in salvation, not in liberty. Perhaps in deference to Jones's objections, the joint committee issued no report and deferred the bill to the judiciary and appropriations committees.[30]

As the days passed, the public warmed to the idea of a single, central state reformatory for white girls. At least, that is how the media portrayed the situation. Letters to editors and articles appeared in major newspapers in support of the plan. The message was clear. A state institution offered the best and most uniform care for wayward children. This was the message in an anonymous article penned for the *Charlotte (N.C.) Sunday Observer*. State institutions that advocated reformatories both for boys and girls provided necessary resources in teaching, discipline, and religious training that local and county reformatories did not. Furthermore, the most important role of the state reformatory, it argued, is to remove the child from a bad environment and give him "new associates." The institution helps to reform a child's character and creates for him a "little world" where the child leaves his past behind him. There, the author asserted, "He has no well meaning and misguided parents close at hand to aggravate his resentment at being sent to the reformatory." A proper system of reformatories is to create four central institutions, the author allowed, "one for white boys, one for negro boys, one for white girls, and one for negro girls." The state, the author contended, can best address the needs of our wayward children "by supporting four great State institutions rather than by supporting many small, poorly equipped, poorly manned, local institutions scattered all over the State."[31]

Few questioned the master narrative that control and care of delinquents rested with the state. The exception was Birdie Dunn, the registered nurse who had come to speak against the bill. She fully opposed state centralization of the care of delinquent girls. In the language of the public health nurse, she argued that communities must handle matters of delinquency at the local level—that is, in the child's community. Her very strong objections to a state reformatory for girls drew from her professional nursing experience in community and public health. In 1902, Dunn had helped establish the North Carolina State Nurses Association. She served as president of the association for twelve years and was active in state and national nursing organizations. She served on the State Board of Health for seventeen years and supported local public health nursing. Her views were common in her profession. Dunn likely had read the work of S. Lillian Clayton, who had published an article in the most recent issue of the *American Journal of Nursing*. In it, Clayton explained the social role of the private duty nurse. The ideal of nursing, Clayton argued, is "not only personal but community service."[32]

Dunn also believed that it was folly to segregate girls from their communities. Most incidents of delinquency stemmed from "some inherited moral degeneracy such as sickness, unemployment, poor housing, careless training, etc." The legislation would affect these victims disproportionately and remove them from their communities. Dunn instead advocated that various public welfare, social work, and health agencies investigate cases of delinquency and address the causes in each individual case. If necessary, the work should be organized under the new State Board of Charities and Public Welfare. State reformatories were "far from ideal," she argued, because they merely segregated victims rather than addressed the root causes of the problem. Let the work be done by medical and nursing professionals and social workers, she argued. She called for the YWCA to build proper social centers for recreation. With such facilities, communities could address girls' problems in their real worlds, rather than segregating them into false "little worlds." "I see no 'fields for harvest' in these 'little worlds,' miserable worlds, that the good women of the State would create," Dunn argued.[33]

This scathing opposition to a state reformatory placed Birdie Dunn in what one historian has described as a "guerrilla war against the intrusion of outside bureaucratic power." Birdie Dunn represents one of those scrappy localists in a fight against an army of well-equipped reformers who argued for social stability and scientific efficiency through state centralization. The conflict between state bureaucratization and local control is known among historians as the "paradox" in southern progressivism. Social reformers desired bureaucratic control and stability in an era of racial and class strife. Prohibition, public health campaigns, child labor reforms, and compulsory school education laws all represented crusading efforts by reformers to create an economically efficient and politically stable southern society, one that emphasized white supremacy and class control. In reaction, localists such as Dunn resisted by adopting a political language suspicious of outsiders and centralized power. Dunn's localism rested in her community-driven nursing experiences. Community nurses, she insisted, utilized the very best modern practices of medicine and science on a case-by-case basis to identify and resolve the root causes of delinquency. Dunn argued that local medical and social work professionals, as well as institutions such as the YWCA, would provide the best care for a community's delinquent girls. Indeed, Dunn was a localist, but a feminist one.[34]

Unlike Dunn, traditional women's club reformers narrowly defined southern womanhood. Women reformers such as Chamberlain sought to further restrict women's behavior by creating a centralized and racialized reformatory steeped in regulations that reflected their own values of white womanhood. Run and legislated by the state, it would streamline every county's process of managing delinquent girls, both white and black. Judges would remand white girls to the reformatory and would send African American girls to county jails or back out to the street. Dunn instead favored a more local approach that rested on a more organic understanding of womanhood. She favored localized solutions to empower local health and social work departments to help girl delinquents. Recreation, healthcare, and individualized attention constituted her solution to delinquency. These are not necessarily feminist solutions, because in her view power rested with the local authorities. Policies favoring recreation and healthcare, depending on the context, might be used as a tool either for restriction or for freedom in behavior. However, her solution allowed for multiple public responses that might include reactionary solutions, but also permitted progressive ideas rooted in inclusive and more fluid definitions of womanhood that might promote feminism and interracial equality.

While Dunn argued in terms that pitted the state against local communities, advocates of the white girls' state reformatory favored a message with a far more powerful subtext, an underlying meaning that rested on racism and white supremacy. White supremacists intended to construct a central state institution that would legislate racist policies by forbidding admission of black girls and women. Reformers such as Chamberlain and McGeachy upheld the purity and potential reclamation of all white women and girls. White women and girls symbolized the purity of the race, and a segregated institution promised to reform them according to their inherent wholesomeness. Critics such as Dunn advocated a localized strategy that did not privilege girls' whiteness or necessarily presume an inherent difference between light-skinned girls and girls of a darker hue. Communities would muster their own resources from local and state sources yet retain local authority. While such localized power would not preclude racialized strategies as solutions, this approach at least did not mandate racial segregation as a single centralized solution. Preserving whiteness simply was not the uppermost priority for modern social work professionals. Even Martha Falconer had classed poor white and black girls together in the category of "menace," and had never made whiteness a major issue. Unlike Falconer, Dunn had never characterized fallen women as a menace or a "war problem." Rather, her public health nursing experiences led her to understand the variety of social pressures that caused ill health and delinquency in communities.

Nevertheless, Chamberlain and her forces prevailed. They had help from the media, for certain. One day before the General Assembly mustered their vote, the editor of the Raleigh (N.C.) *News and Observer* issued a glowing letter of support for the clubwomen and the state reformatory. In his missive, the editor upheld the purity of white womanhood, the sacrifice of the noble clubwomen, and the melodramatic narrative. In defense of the reformatory bill, he employed the well-worn victimization story. "A man with horried [*sic*] lies," he argued, "wins a woman's love, seduces and causes her to fall from the pedestal of beautiful, noble womanhood." Race saturated his rhetoric, as white women not only shared efforts to "elevate and ennoble humanity," but also had histori-cally provided service to the noblest of all causes, the Confederacy. Legislators must vote for the bill. "How can our State forget their consecrated service to the boys in gray," he argued, "when the red waves of battle dashed over our war-blasted and bleeding country?" It is perhaps this language that most appealed to the legislators. The final vote came on March 7, the last day of the session. According to one source, it was a stormy winter night. Representatives began to leave the chamber at midnight, the legal hour for adjournment. The Speaker of the House, Walter "Pete" Murphy, made good his promise to Chamberlain to get the bills passed. Just as the bells began to toll, Murphy ordered the sergeant at arms to spin back the clock fifteen minutes and to lock the doors. He barely maintained a quorum, at sixty-one members. Both the reformatory bill and the $25,000 appropriation passed unanimously on the final vote, sixty-two to zero. The State Board of Charities and Public Welfare bill passed, too. Chamberlain and her ladies, present that night, were victorious. White womanhood pre-vailed, and everyone went home.[35]

2

In Defense of the Nation

SYPHILIS, NORTH CAROLINA'S "GIRL PROBLEM," AND WORLD WAR I

WHILE REFORMERS CELEBRATED PASSAGE of legislation to establish a white girls' home, the funding that actually built the campus came from a very different source. The federal government, in the context of controlling venereal disease in World War I, provided substantial monies to build and open Samarcand Manor on the condition that it also house adult women charged with prostitution. During World War I, the State Board of Health feared the spread of prostitution in the proximity of military training camps. The board launched a campaign to protect soldiers from venereal disease by controlling the sexuality and freedom of movement of girls and young women. North Carolina had a "girl problem," state authorities and officials insisted. To solve it, they constructed state policies based on Victorian ideals of morality. In these terms, single women fell into only two categories: virgins and fallen women. Legislation and rules endorsed by the State Board of Health characterized delinquent girls not as wayward juveniles (a third category inconceivable by Victorian standards), but as potential prostitutes who required segregation, quarantine, and study to protect potential enlisted men. The "girl problem" existed throughout the nation, but it was particularly acute in the South because adolescent girls who exhibited sexual self-determination not only challenged the gender hierarchy but also threatened white supremacist associations of chastity and whiteness.[1]

The greatest irony lay in the founding of Samarcand. While its spirit rested in the redemption of white girls, the financing drew from federal policies to criminalize prostitution. In 1919, reformers' ideals of southern white womanhood clashed with ideals of Victorian womanhood enforced by the U.S. government in policies toward venereal disease control. Reformers viewed the white girls at Samarcand as redeemable, the feminine representations of white supremacy. The U.S. Army, by contrast, viewed them as diseased prostitutes, the worst of Victorian stereotypes. Any federal financing rested upon the contingency that the North Carolina General Assembly pass laws to control prostitution and thus address the "girl problem." This chapter examines the role of the North Carolina State Board of Health in enforcing the policies of the U.S. Army. Endorsed by a World War I-era federal campaign to protect soldiers from venereal disease, North Carolina passed gender- and class-based antiprostitution and venereal disease control laws that reinforced a sexual double standard, one that acknowledged the sexual impulses of men while repressively punishing women for having the same impulses. With no place else to put convicted prostitutes, the state petitioned and received federal funds to open Samarcand Manor as a school to serve delinquent girls, but also to serve as a quasi-prison for adult women charged with prostitution. These developments attached an early stigma to Samarcand and its female residents. As a result of this paradoxical mission, Samarcand Manor represented a clash of two ideals of womanhood; the reformatory came to be etched in the public mind as a place where white adolescents associated with hardened prostitutes.[2]

Historians have shown that society attaches social beliefs to venereal disease. Before the twentieth century, sexually transmitted diseases were viewed as "carnal scourges" representing sin and transgression. Once, society had viewed venereal disease as punishment for sin. But as scientists learned more about bacteriology, the meanings changed. Progressive-era reformers viewed venereal disease as a family based problem that required public solutions. Some reformers saw syphilis and gonorrhea as sexually communicable diseases that required rational and efficient techniques, including education and treatment, to control it. Others argued that the best form of eradication was to reinforce early stigmas associated with sexually transmitted diseases. These reformers hoped that the fear generated by these stigmas would deter men and women from dangerous behavior associated with infection. A disease of disgrace and dishonor, venereal disease became a moral sign of decay that threatened a rigidly moral sexual order. These moral codes prescribed the sanctity of the family and women's inherent role as mothers. The ideal woman was chaste, virginal, and innocent, and the ideal man was the family provider. Venereal disease threatened these sentimental ideals of clean and pure husbands and wives by drawing attention to extramarital sex and by disrupting the health and sanctity of the family. Furthermore, it upended gender roles. A man with venereal disease threatened his family's safety, and an infected woman missed her natural calling as a mother. In this light, public fears and stigmas associated with venereal disease died hard.[3]

Venereal disease terrified progressive reformers not only for its physical manifestations, but also for the threat it posed to the moral purity of the manly American. During World War I, standards of American manhood were in flux. Woodrow Wilson and other prowar reformers embarked on a moral purity campaign designed to shore up the health, efficiency, and manliness of the nation's fighting forces. It was an ambitious plan. Soldiers had developed a less than pure reputation. Since the American Revolution, the Civil War, and conflicts along the Mexican border, the popular mind had linked soldiers and military encampments with immorality, vice, theft, and profanity. World War I-era progressives planned to change that reputation. The Commission on Training Camp Activities (CTCA), established by the War Department in April 1917, promoted a form of cultural nationalism intended to remake American manhood into a national standard that, historian Nancy Bristow argues, defined manhood in middle-class terms. Under the aegis of the War Department, the new American man would be sexually pure, self-controlled, physically fit, and ready to accept broad responsibilities at home and abroad. Bristow demonstrates that the new manhood accompanied ideas about womanhood as well. Reformers generally embraced women's expansion of the domestic sphere into public affairs. Yet CTCA posters, advertisements, and policies represented women as sexualized objects, either as chaste virgins or fallen women. The dichotomy ensured that most CTCA and War Department policies defined women in terms of their relationship to men. In other words, women's bodies, leisure, and health were defined only insomuch as they would be regulated to facilitate the physical and moral fitness of men.[4]

Thus, gender biases shaped the CTCA plan for eliminating venereal disease (one of its chief goals) and red-light districts (the chief target). The CTCA identified prostitutes as the source of infection of young, innocent men. It was no easy task to wage a battle against the nation's red-light districts. Restricted districts, though always controversial, operated in many towns and cities as a way to segregate impure women from pure ones and to control venereal disease. The CTCA joined forces with antiprostitution social purity crusaders to close many of the districts and to encourage communities to pass strict antivice legislation. The CTCA achieved broad success thanks to its federal enforcement powers and the economic incentives it wielded via its control over the location of military camps. Yet it wielded little power over the prostitutes. Once districts closed, women moved outside the city or near encampments and men continued to meet them. In 1918 the CTCA changed its direction by shifting its attention from the red-light districts to the prostitutes themselves. To achieve this new goal, the CTCA embarked upon repressive efforts of eliminating prostitutes by harassment, arrest, quarantine, forced treatment, and detention in camps, houses, and reformatories. Many communities suspended women's rights in an effort to arrest and jail not only prostitutes but also any women or young girls suspected of immoral sexual activity. To address this "girl problem," the CTCA created the Committee on Protective Work for Girls (CPWG).[5]

The campaign intensified in 1918 when Congress passed legislation that strengthened the CTCA. On July 9, 1918, Congress passed the Chamberlain-Kahn Act. The act created the Interdepartmental Social Hygiene Board (ISHB) and provided the board funds to distribute to the states for the purpose of controlling the spread of venereal disease. Congress intended the act to protect the "man power" of the United States by protecting soldiers from diseased civilians. In its manual, the board regarded prostitutes, "both male and female," as a national liability and blamed them for the spread of venereal disease. Every soldier with venereal disease, the army concluded, had contracted his infection from a civilian source. Thus, the national government was obliged to help the states prevent, treat, and control such infections. The act specifically empowered the secretary of war and the secretary of the navy to assist states in identifying, isolating, quarantining, treating, and institutionalizing any civilian who presented a threat to the armed forces.[6]

Although the Chamberlain-Kahn Act did not specifically single out women as the objects of control, the ISHB manual discussed at length the problem of delinquent women and girls, or "camp-girls," who settled near military training camps throughout the nation. To manage these women, the War Department had begun to create detention houses and detention hospitals "where all women and girls (except hardened cases) who are arrested may be held while awaiting trial, to be studied and treated medically." An act of Congress passed that year made it a federal crime to prostitute within five miles of the training camps. Conviction brought a $1,000 fine or imprisonment for one year. President Wilson earmarked funds for reformatories for the custody and rehabilitation of girls and women "who proved to be a menace to the health and morals of the men in training." Some authorities briefly considered building concentration camp-style detention centers in the four quadrants of the nation, but they abandoned the idea due to "practical difficulties of execution." Finally, the board concluded that a national system of reformatories was prohibitively expensive and settled on a plan to aid the states in building reformatories and industrial schools of their own.[7]

Many North Carolina reformers, particularly physicians and legislators, embraced the federal campaign and its effort to remake American manhood. In 1919, Dr. James A. Keiger, director of the State Board of Health, urged the North Carolina General Assembly to join a nationwide campaign against venereal disease because "in the Army, the physical effects of the venereal diseases were more disastrous than wounds received in battle." Army doctors had examined 10 million men, "American citizens, drawn indiscriminately from all social strata." Each man stripped and succumbed to a thorough examination. "The amount of venereal disease that was found was simply astounding," Keiger reported. In the U.S. Army, over eighty thousand cases of venereal disease were recorded between September 1917 and June 1918. Incidences of venereal disease were highest in newly enlisted men; 85 percent came from the civilian population and only 15 percent were contracted after enlistment. Army officials insisted that venereal disease kept more men from the trenches in Europe than battle wounds.[8] While the army's rate of venereal infection was proportionately low, the surgeon general reported that the prevalence of venereal disease was nearly five times all other serious contagions combined.

Venereal disease, chiefly syphilis and gonorrhea, presented such a problem that officers fantasized how to be rid of it. Keiger quoted to legislators the words of the surgeon general of the U.S. Army, who had concocted his own fantasy about the eradication of venereal disease: "If the medical department of the army had a choice presented to it, say if some man came with a wand, and it was demonstrated that with this wand every wounded man could be gotten back into the line [of duty] at the end of the second day, with his wound cured; and another course were presented by which all venereal diseases could be eradicated from the army, and our choice were given . . . I think there would be very little hesitation on the part of our department in choosing the eradication of venereal diseases."[9] It is clear that army doctors considered these parasitic diseases debilitating, difficult to treat, and logistically problematic for maintaining a healthy force. Yet this man-wizard imagery offers further insight into prevailing attitudes about male sexuality. The quote presumes the sexual impulses of men and addresses venereal disease not as a moral character fault in them but as an expected though undesirable consequence of war. The surgeon general's comments also identified the problem in terms of class, suggesting only enlisted men, not officers, were susceptible to venereal disease. Thus, men of higher standing (the man with a wand, presumably a physician) should help uplift the rank and file in an effort to combat this unintended enemy as if it were any other.

State Board of Health reports and letters indicate that in 1919 venereal diseases raged through North Carolina. The state ranked tenth in a study of cases of venereal disease per thousand soldiers enlisted, exceeded only by seven of its southern sister states, Oklahoma, and West Virginia. In North Carolina, Keiger estimated that syphilis infected 5 percent (that is, 12,173 individuals) of North Carolina men aged twenty to thirty-four. He estimated the incidence of gonorrhea in North Carolina men at 5 times greater. Though Keiger admitted he had no precise statistics for the incidence of venereal disease in the female population, he estimated the numbers at "1% of the female population," or 2,672 cases.[10]

Prior to the federal legislation, North Carolina cities had handled venereal disease in city ordinances. Some practices acknowledged anonymity regardless of gender. Numerous cities had established clinics to handle voluntary treatment of venereal disease. In Asheville, North Carolina Health Department physicians offered care and anonymity to anyone who willingly pursued treatment. One prominent physician there, Dr. Carl D. Reynolds, deemphasized prostitution as the cause and stressed that "any man or woman should be treated not as a criminal but as one with a disease." Further, he supported sex education and laws that raised the age of consent to twenty-one. Greensboro, in Guilford County, had clinics too. The county also had adopted a "morals ordinance" that outlawed bawdy houses.[11]

Local public health campaigns in cities such as Asheville, Greensboro, and Charlotte reflected the general characteristics of southern social policy at the turn of the twentieth century. Southern localities possessed considerable autonomy in areas of social policy, such as public health, vice control, and public school education. State boards of public health intervened only during emergencies and epidemics. Before 1919, state bureaucrats generally lacked any coercive powers and simply relied on voluntary cooperation by county health officers. Sometimes county health officials and local physicians, antagonized by medical professionalization and state intervention, refused to cooperate. The medical industry, some people argued, had invented the idea of germ theory and public health disbursal of drugs and vaccines to generate profit. In 1913 one North Carolinian declared that the science behind microbes allowed physicians to "multiply serums, add to the fears, subtract you from your money and divide all the profits among the doctors." Other North Carolinians simply refused to report incidents of infectious diseases among their neighbors for fear of making enemies.[12]

Did local resistance to state centralization represent a lackadaisical attitude toward public health? Perhaps it did in some locales. But many cities in North Carolina mounted aggressive campaigns to fight the spread of infectious diseases. These campaigns reflected autonomous efforts to control epidemics, but also the efforts of local health authorities to seek advice and counsel from the state.

The city of Charlotte aggressively fought venereal disease. Less sympathetic to anonymity of venereal disease patients, Charlotte authorities publicized names of the diseased for quarantine purposes, regardless of gender. In March 1918, a few months before the passage of the Chamberlain-Kahn Act by Congress, the Charlotte Board of Commissioners established its own "Ordinance for the Control of Venereal Disease." This ordinance made it unlawful for one person to expose another to gonorrhea, syphilis, or "chancroid infection," required individuals to seek treatment, and allowed prosecution of any individual who failed to comply with a licensed physician's treatment regimen. The ordinance further empowered the health officer to quarantine known infected persons and placard their residences to announce the presence of the disease. In addition, the ordinance required that only licensed druggists could disburse medication. The druggists were to keep records of the date of sale, kind and quantity of medicine sold, and name and address of the person making the purchase. All inmates of the city jail could be examined. Finally, the ordinance made it unlawful for any person "in the infective stage" to handle "public drinking glasses or vessels" or to work in any place where food or drink was handled.[13]

With this ordinance, Charlotte health authorities intended to find all cases of venereal disease, identify them, and control the danger of infection to the public, regardless of the sex of the infected person. For example, the ordinance criminalized any person who spread the infection or refused medial treatment or quarantine. Additionally, the ordinance protected consumers from all individuals with venereal disease in certain workplaces and restaurants. It is not possible to fully examine the use of the ordinance in terms of gender, because the state law superseded it a year later. However, the ordinance had teeth and, according to its wording, would have required the city to control infection not only by prostitutes but also by the married "johns" and potential enlisted soldiers that might have spread the infection themselves.[14]

Local concern about venereal disease was really rather new. The public health laws of 1911 and 1913 that allowed communities to police epidemics and quarantine the sick listed numerous contagious diseases, including whooping cough, smallpox, dysentery, and influenza, but made no mention of venereal disease. Cities' interest in adding venereal disease to the list of controllable contagions reflected both progressive impulses of rational and scientific attitudes toward public health and social order and scientific advances of the early twentieth century in combating the disease. Though syphilis and gonorrhea were sexually transmitted contagions and thus a different sort of public health menace than airborne diseases, cities such as Charlotte applied similar regulations for reporting, isolating, and quarantining the sick.[15]

Cities fought venereal disease in much the same way as they battled other infectious disease epidemics in their communities. When Spanish influenza struck North Carolina in 1918, local city and health officials mobilized to warn their communities. Depending on the severity of the epidemic, they employed a wide spectrum of tactics, including quarantine, temporary restrictions, fines, and awareness campaigns. One community required reports of every influenza case and further prohibited social or "business" visits between families "except when such person or family is in distress or need due to sickness or other affliction." The code, passed October 18, 1918, would remain in effect until revoked by the board of directors. The city would charge offenders with a misdemeanor and a fine "not to exceed $5.00 for each offense." On October 31, the Columbus County Board of Health required bottling plants to undergo sanitary inspection, banned "public drinking cups," and ordered individual serving "cups, saucers or cones" for all servings of soft drinks and ice cream. In their most radical move, they decided to close all churches and schools for nine days.[16]

During the crisis, the State Health Department served chiefly as an advisory board to panicked officials across North Carolina. State law granted local health boards significant protection and decision-making power in the event of epidemics. Local health authorities determined the restrictions and duration of these restrictions on the local populace. The State Board of Health only established minimum quarantine regulations and gathered information from county officers, who were required to report all incidences of contagious diseases. State health authorities investigated the cases and issued recommendations and advice to county health departments. The state stepped in only when there was no regularly organized local board of health. During the influenza crisis, health officers throughout the state wrote the State Health Department to report their local activities or to request advice. Communities acted on suggestions to organize local health committees and volunteer groups.[17]

In batting epidemics, local authorities drafted policies that restricted behavior and movement of residents that were relatively gender neutral. Practically speaking, syphilis and gonorrhea were gender-blind diseases that infected both men and women. Prostitutes threatened to spread the disease to unsuspecting soldiers. However, an equal threat for many local health officials were the infected men who could spread the disease to unsuspecting wives. Communities possessed considerable autonomy in their battles against contagious diseases. While they often requested advice, they constructed their own defenses as the crisis unfolded in their communities. State law provided a minimum of oversight, but it largely affirmed the authority and autonomy of local officials. What is interesting is that some of the larger communities, including Asheville and Charlotte, approached the control of venereal disease in a more or less nongendered manner. While some authorities noted that prostitutes required control, they argued that the criminalization of prostitution was beside the point. Instead, they prioritized a rational system of reporting, isolating, and treating contagious diseases to protect the public health.

This local approach did not suit the U.S. armed forces. While cities pursued venereal disease control as they did most other communicable epidemics that threatened men, women, and children alike, the military viewed syphilis chiefly as a threat to the fitness of soldiers. The army began its campaign just eleven days after the American declaration of war. On April 17, 1917, the War Department established the Commission on Training Camp Activities to take charge of soldiers' recreation and to control the red-light districts that sprouted around military training camps. The commission's widely distributed *Keeping Fit to Fight* pamphlet encouraged soldiers to remain chaste for the preservation of their manhood and virility, pitched the horrific effects of gonorrhea and syphilis not only to the soldier but to his family at home, and stressed the threat of disease to military efficiency. One social hygienist considered prostitutes to be worse than the enemy: "It is generally recognized that a bad and diseased woman can do more harm than any German fleet of airplanes that has yet passed over London. One woman of such character as effectually destroys a soldier as a German gun would, and more so. A German gun might do much more harm to men in front of it, but it would not leave the wounded as a menace to his associates, as is a man who suffers from 'social diseases.'"[18]

Language that associated prostitution with German bombers died hard. North Carolina legislators, strongly encouraged by the U.S. Interdepartmental Social Hygiene Board (ISHB) and the military authorities that staffed the agency, moved to pass laws that defined venereal disease as a prostitution problem. In North Carolina, the ISHB enlisted the state to usurp the communities' local boards of health. In 1919, the state of North Carolina passed legislation that rescinded some of the control of local boards of health. Legislators and Board of Health officers used city ordinances as models, but also vetted the bills through the federal ISHB. In fact, some legislation was standardized by the Interdepartmental Social Hygiene Board itself. The board consisted of the secretaries of the navy, army, and treasury and the surgeon generals of the navy, army, and the U.S. Public Health Service. If the occupation of the board members is any indication, then clearly military priorities predominated. Uniformed officers in distant halls, not local progressive reformers, bore the greatest influence in drafting North Carolina's state venereal disease legislation.[19]

A full-page advertisement in Raleigh's *News and Observer* evidenced the alliance between the State Board of Health in Raleigh and the U.S. Public Health Service in Washington, D.C. The ad warned North Carolina readers to prepare for an assault on peacetime prostitution. "Your whole community will be at the station 'when the boys come marching home,'" the ad cautioned, "are you making sure that the profiteers of vice are not planning to take advantage of the days of festivity to dishonor them before they get settled again in the normal ways of life?" The threat required more than mere police enforcement against red-light districts. The state must pass legislation to eliminate the vice. "Diseased prostitutes are the most dangerous carriers," the ad warned, "they must be quarantined and the community safeguarded against their return as prostitutes first by means of permanent segregation of the feebleminded and, second, by medical treatment and industrial education for the others." Ideally, state hospitals and reformatories for women and girls would serve the purpose of permanent segregation. Unlike the ordinances previously passed by local boards of health, nothing in the ad mentioned anything about quarantine or treatment of the pimps and johns.[20]

Apparently the ad succeeded in persuading legislators that prostitution was the source of venereal infection. In 1919 the General Assembly passed its first significant legislation against prostitution to control venereal disease. The four acts that composed the state venereal disease legislation directly targeted prostitutes and required no significant oversight of the men who frequented them. The first act, drawn from Guilford County, prohibited bawdy houses throughout the entire state. The second act made it unlawful for any person to aid or abet prostitution. The third act allowed for the state to quarantine prostitutes and to placard their homes at the discretion of the health officer. The law required the treatment of prostitutes at county expense to make them noninfectious. One enigmatic provision allowed the Board of Health to employ three plainclothes men, such as Special Agent "X," "to find and deal with prostitutes." Finally, the fourth act required physicians and druggists to report all incidents of venereal diseases and all drug sales intended for treating the afflictions, but unlike the Charlotte ordinance, it allowed anonymity of patients. Indeed, many of the provisions echoed the Charlotte ordinance, but with significant modifications that specified prostitutes and prostitution as the key focus of control and investigation. Later, the Board of Health suggested examination of all pimps and sex offenders arrested by county police, but it remained clear that the thrust of the law was aimed at women and girls.[21]

The federal government's Victorian ideal of womanhood offered no middle ground between the categories of virgin and fallen woman. Authorities defined a fallen woman as a prostitute. Broadly defined, a prostitute was a woman who engaged in sex as part of a cash transaction or a sexually active girl who might later gravitate toward the profession. The campaign against prostitutes targeted the working class. Sexually active girls, also known as "charity girls," posed a potentially greater problem than the hardened prostitute. The term "charity girl" originally applied to young working-class women who used sexual relations to barter for gifts and amusements. During World War I, the term came to apply to any girl who engaged in sex with soldiers free of charge. Middle-class girls adopted some of these courtship practices. They described their exchanges with

boys and men not in terms of "charity" but as "treating." In fact, the "girl problem" was especially insidious because it potentially included women of middle-
class backgrounds. The flapper, a popular stereotype that depicted young girls
as promiscuous, rebellious, and sexual, was especially dangerous. Only a fine
line of behavior separated middle-class flappers from working-class "charity
girls." Though middle-class girls did not usually suffer reputations as hardened
prostitutes, they did risk charges of being delinquent, defective, or feebleminded. Involuntary detention, quarantine, medical examination, and arrest
were common ways for social hygiene reformers to control suspect women. The
moral double standard operated with impunity. Laws and health regulations
did not establish such punishments for men who defied the sexual codes.[22]

One state officer frustrated by the new but one-sided venereal disease control laws was Dr. Millard Knowlton, the new chief of the Bureau of Venereal
Disease, a division of the State Board of Health. Knowlton rejected the sexual
double standard and knew that an effective quarantine policy required immediate and thorough action regardless of a person's sex or identity. Quarantine enforcement was critical to control any epidemic. In October 1919, Knowlton advised a Kinston health officer to arrest and quarantine anyone, male or female,
who refused treatment for the disease. The local officer refused Knowlton's
order and complained to Knowlton's superiors at the State Board of Health.
For some reason, the Kinston officer hesitated to arrest and embarrass a man
of prominence. The physicians on the State Board of Health advised the Kinston health officer to disregard Knowlton's command. To Knowlton's dismay,
the sexual double standard, reinforced by his superiors at the State Board of
Health, prevailed.[23]

Rather than risk embarrassing men of prominence, the State Board of
Health focused its attention on the efforts by police departments and local
courts to curtail prostitution. The board, lacking police jurisdiction, secretly
employed private detectives to monitor local red-light districts and to report
on the activities of known pimps, johns, and prostitutes. One detective, known
as Special Agent X, crossed the state from one red-light district to the next. His
modus operandi was to locate taxi drivers and other city "chauffeurs" or to sur-
reptitiously follow a john, who would deliver him to waiting brothels and pimps.
Sometimes these men were the most eminent in town. One time, in Kinston,
he followed a john into a notorious brothel called Flommie Gulley's on Shine
Street. Once inside, Agent X realized that the john was the chief of police.[24]

Agent X reported the chief for his violation of duty and recommended to his supervisor, a physician, the removal (not the arrest) of the police chief and the reorganization of the police department. Throughout the summer and into the fall of 1919, Agent X continued his detective work in twenty-seven cities across the state. In July he visited Durham, High Point, and Farmville. In Greensboro that month he found "Conditions bad. Hotels immoral," and "houses of prostitution existing." Raleigh received his visits several times, and by the end of the summer he noted, "conditions improving steadily." In October he investigated several cities, including Greenville, where prostitution flourished (he discovered three "open red light houses") and the police turned a blind eye. Usually the police actively assisted his investigations. He noted that the police chief in Monroe, North Carolina, was "an efficient officer," though the judge there usually rendered his verdicts "in favor of his lawyer friends." Other judges went easy on perpetrators. Agent X found that the judge in Greensboro tended toward leniency with women arrested for prostitution. When Agent X first visited in July, he found that the judge merely asked the girls to leave the city. By the agent's second visit in October, the judge had taken the bureau's recommendation to more heavily punish the women by sending them to a juvenile delinquency reformatory called Samarcand Manor.[25]

In such an environment, a man's affluence virtually ensured the protection of his identity. Following the October 1919 incident in Kinston, the State Board of Health took proactive measures to protect the identity of prominent men afflicted with venereal diseases. In December 1919, board members met to suggest further rules and regulations to protect the identity of patients. The original legislation required that anyone, male or female, seek treatment if they suspected an infection. However, the board recommended privacy for any man who discontinued treatment. Authorities were to "take such other steps as may be deemed necessary to properly safeguard the rights of the patient and at the same time adequately protect the public health," but only at the discretion of the local public health officer. The board further indicated that all reports of venereal disease should be regarded as confidential and records should be closed to public inspection. At a physician's discretion, reports of disease should be kept by number, rather than in the name of the patient. The board would make "every reasonable effort consistent with the protection of [the public health] to keep secret the identity of persons affected by venereal disease control measures." By law, then, alleged prostitutes who refused treat-

ment were to be publicly embarrassed with seven-by-eleven-inch yellow plac-
ards bearing their names and type of disease, but prominent men who refused
treatment might seek anonymity at the discretion of the local health officer.[26]
In other words, the law applied to both men and women of all classes, but a
man who might seek discretion with local authorities had the support of the
State Board of Health.

In 1919, medical professionals understood that the standard treatment for
syphilis imposed upon women was deadly. The oft-quoted phrase "if the disease
doesn't kill you, then the cure surely will" fittingly describes the treatment at
that time. Standard technique included some combination of injection, inunc-
tion, and intravenous application of arsenic and the heavy metals mercury and
bismuth. Physicians fully recognized the danger in such treatments. "It is a
well known fact," stated a 1931 report of the State Venereal Disease Advisory
Committee, "that no drug used in the treatment of syphilis is absolutely safe."
In fact, reports advised very careful examination and diagnosis of patients,
because in some cases "intensive treatment," one physician warned, "would
be tantamount to malpractice." For a patient with advanced heart disease, he
warned, "the results of too intensive treatment may be disastrous." The State
Board of Health warned that the initial doses of arsenic should be small and administered "*slowly*" to "obviate probable severe therapeutic shocks." Physicians
were to inject arsenic-based drugs but take care not to agitate solutions. Any
vigorous shaking could oxidize the solution, thus rendering it toxic. "At the first
symptoms of arsenic poisoning," the board warned, treatment must end. More
recommendations followed for the physician faced with such an unfortunate
patient.[27]

The State Board of Health treated prostitutes by rubbing mercury on the
skin. State physicians did not recommend the administration of mercury by
mouth. Doctors might administer insoluble mercury intravenously, but this
treatment was, according to the state health records, painful, "cannot be controlled, and its action is uncertain, and may even be dangerous." Soluble mercury was safer, but daily rubbing of mercury on the skin was preferable. The
State Board of Health warned doctors to follow very careful dosage instructions
because this form of treatment, state health records vaguely warned, "carries
with it the danger of betrayal [death] of the patient." In addition, the advisory board recommended injections of bismuth and mercury with a medium-
length needle to penetrate the muscle. Above all, physicians should take care

to avoid injection into any vein because "even small amounts of certain bismuth preparation, intravenously," the advisory board warned, "may be fatal, and any oily solution intravenously gives rise to the danger of embolism." Thus, syphilis treatment required intensive scrutiny and careful supervision of patients, not only to render them noninfectious but also to prevent poisoning or, in the words of the advisory committee, "betrayal of the patient." The state's treatment policies reflected profound gender and class implications. While a prominent man could refuse this very dangerous treatment, the State Board of Health provided no such favor to a woman, particularly the prostitute or poor "charity girl."[28]

Thus, the state preferred controlled examination, isolation, and treatment, especially of female patients. It was no easy task to force treatment on female patients. So the state drew upon the Chamberlain-Kahn Act for help. In 1919 and 1920, the Interdepartmental Social Hygiene Board directed funds to a state reformatory called Samarcand Manor. Samarcand, a state juvenile delinquency institution for white girls, badly needed money to remain solvent. The Samarcand superintendent submitted applications at least twice to the federal board. The January 1920 application stated that of 146 inmates, the average number infected with venereal disease was 109. (This number is likely inflated, because other records indicate that usually no more than 30 percent of girls at Samarcand had venereal disease.) Furthermore, the application mentioned the "nearby" military camps. According to the application, the closest to Samarcand was "Camp Bragg," thirty miles away. As a result, the federal board appropriated to Samarcand an initial $10,000 in 1919 and $20,000 in 1920, with strict instructions that the funds be used specifically "for the care of civilian persons whose detention, isolation, quarantine or commitment to institutions has been found necessary for the protection of the armed military and naval forces of the United States."[29]

By the spring of 1920, the strict mandate tied to the national funding had begun to conflict with the institution's primary mission of serving as a correctional center for juveniles, ages nine to sixteen. Most of the girls were young children, the key population target of the institution. Sentences were indeterminate but by law were not to exceed three years. Many girls were committed by parents and community authorities for minor infractions. Some were orphans with no criminal record, but who possessed a family history of ne'er-do-wells. Others were infected with venereal diseases. The campus housed an infirmary staffed by a male physician and a woman nurse. The institution provided dorm living, education, and vocational training to set the girls straight before ultimately sending them back to their communities. Older girls and young women were not among the target population. But after the federal government began its appropriations in 1919, older girls and young women were sent to the institution. Special Agent X noted several alleged prostitutes who were sent to Samarcand from Greensboro, Rockingham, Rocky Mount, and Winston-Salem in late 1919. In January of 1920, the commissioner of the Board of Welfare, R. F. Beasley, admonished the Samarcand authorities for turning away a girl "because she was over sixteen." Apparently the girl's parents and the local court authorities wanted her committed due to her "waywardness, vagrancy, and lewd conduct." In his letter to a Samarcand board member, Commissioner Beasley advised that the superintendent devise other reasoning for refusing admittance to older girls and women while the federal government sent funding. "In fact," Beasley asserted, "this help is predicated rather upon the idea of the isolation of the very type of woman that was refused entrance." Samarcand Manor, a juvenile delinquency institution, had become a reformatory for prostitutes as well.[30]

From its very origin, Samarcand Manor served contradictory purposes. It housed adolescent white girls in need of redemption as southern ladies and also adult prostitutes in need of punishment, treatment, and control. Two competing definitions of womanhood, one based on southern whiteness and the other rooted in Victorian morality, each viewed sexually active women as being aberrant. However, the first presumed that white girls could be retrained as southern ladies. The other presumed that adult women, beset with mental deficiencies, had fallen irredeemably to hardened and diseased living. Southern white womanhood clearly possessed elements, such as the celebration of chastity, related to its Victorian counterpart. However, in white supremacist fashion, the southern view emphasized that white girls, by the superior nature of their race, could be redeemed. Perhaps most confusing and contradictory was that the county courts committed white adult women to Samarcand Manor. It seemed that hope remained among North Carolina authorities that adult, white prostitutes also might be saved, a significant inconsistency in the federal government's Victorian ideal.

North Carolina filed no application for funds to control African American girls. Scholarship shows that this lack of attention toward African American girls was far from benevolent. Deep-seated racial prejudices in North Carolina and in the nation prevented authorities from viewing African American girls as worthy of saving. No state reformatory existed for African American girls in 1919 because many southern whites viewed this population as innately immoral, unreformed, and controllable only through the criminal system. These preconceptions persisted despite venereal disease rates in African American girls that approximated those of white girls. Even clubwomen of both races found the stereotypes difficult to fight. Beginning in 1925, African American and white clubwomen financed a privately held reformatory for African American girls called the North Carolina Industrial Home for Colored Girls, also known as the Efland Home. Despite intense lobbying by African American and white women reformers, the North Carolina General Assembly provided operational funds of only $1,400 to $2,000 each year. All other funding came from private sources. The school paroled more than two hundred girls before it closed its doors in 1939 due to inadequate financial support.[31]

The Chamberlain-Kahn monies provided to open Samarcand Manor produced conflicting messages about the purpose of the institution and the composition of its inmates that resonated for years. The year 1919 marked North Carolina's first state legislation on venereal disease. Reformers founded Samarcand Manor on the myth of southern white womanhood. However, the state turned to federal financing tied to the control of prostitutes, viewed as the source of the venereal disease scourge. Victorian ideas of womanhood that viewed all sexually active girls regardless of race as fallen preempted the white supremacists' plan. Furthermore, the state passed unprecedented legislation enforcing the Victorian ideal that tied venereal disease to the moral issue of prostitution, thus complicating eradication. Furthermore, the gender-biased laws that provided anonymity for male patients contributed to perceptions that sexualized girls and young women and linked their deviant behavior, however benign, to prostitution and disease. These developments served as the seed of the stigma that beset Samarcand Manor and its residents for the next twenty years.

3

How to Make Bad Girls Good

DISCIPLINE AND RESISTANCE INSIDE A
GIRLS' REFORMATORY, 1918–1925

SHE HAD KITCHEN DUTY, AND this morning she had awoken late. Without bothering to don her shoes, she slipped into the kitchen and found two pails of food waste awaiting the hogs. Old bread, rotted fruit, hominy, these hogs got the good stuff. Mrs. MacNaughton had big plans for them at the state fair in Raleigh. The trek to the hog pen generally was not a dangerous one for a barefoot girl. Usually, the only risk was the low-growing cacti that sprouted in the sand and might spear a little girl's toes. But this morning greater dangers lurked. Here sooii! The slop fell into the trough. There, pinned to the hog pen, was a note. Three men, it said, had hidden in the pines four hundred feet away and were ready to help any girl make her escape. She rushed madly from the barn to the Manor to alert Mrs. MacNaughton. It was a common occurrence. Girls rumored delinquent aroused the curiosity of mischievous young men. The staff was ready. Three young college teachers dressed in girls' bloomer suits went out to meet the trespassers. The men ran toward them with outstretched arms until they realized the young ladies were not inmates, and then turned and fled "as men never fled before." The teachers, bearing revolvers, pursued them and fired shots in the air, which only served to quicken their pace. Later, Agnes MacNaughton, superintendent of the State Home and Industrial School for Girls, also known as Samarcand Manor, provided comment to the press. "Our most serious trouble is from the outside," she confessed, "and as soon as prowling around our grounds with evil intent is made a felony quite a load will be taken from our shoulders."[1]

The word was out that bad girls lived at Samarcand Manor. The press knew it, the state knew it, the federal government knew it, and even the bad boys knew it. It was a difficult stereotype to overturn. Samarcand Manor, established in 1918, served as North Carolina's only juvenile delinquency center for white girls. Girls came to Samarcand with social maladjustments, venereal diseases, felony convictions, prostitution arrests, and even lesser misdemeanors like vagrancy that might not warrant such excitement. They were cotton mill workers, girls from the streets, and sometimes both. Some had no record at all. "Truants-on-the-danger-line, such as the little girl of 10," described one journalist in the *Orphan's Friend and Masonic Journal*—"have been saved untarnished; others came from dangerous homes, others were of the half-world, and a considerable percentage of the latter were infected when they arrived." MacNaughton's public statements and correspondence conveyed every image of successful reform, but the reality belied the rhetoric. The conventions of white womanhood, specifying purity and submissiveness, had created a predicament for administrators. These were girls who had witnessed, suffered from, or participated in nearly every social transgression. The state had charged Samarcand with the mission of transforming them into genteel southern ladies. In some girls it met with some success, but in others it failed, unable to erase the most incorrigible girls' troubled pasts. Thus, administrators at Samarcand constructed a polarized system of governance, defined chiefly as a publicly lauded "honor system" and a more secretive "probationary" system, to create an appearance that the institution succeeded in reforming bad girls into good southern ladies.[2]

White supremacist imagery lay at the core of southern whites' desire for a racially segregated girls' institution. In the U.S. South, the ideal of white womanhood rested at the heart of a racial hierarchy. The conventions of the day characterized white women as virtuous, pure, genteel, and the exclusive object of white men's affections and desires. Those same conventions restricted African American women from claiming ladyhood. The sexual and racial relations of the South created a multidimensional double standard where white women represented gentility and purity and black women symbolized sexuality. White men could engage in sexual liberties across racial lines and black men could not. This confluence of racial and gendered stereotypes thus reinforced four tiers. The first elevated white men, the second degraded black men, and the last two treated all women as objects—one set viewed as privileged and pure, the other set defined as sexual and servile. To maintain this multitiered fiction, southern whites needed to elevate the lowest class of white girls to the status of ladyhood, that tier of the privileged and pure, if only in name.[3]

It almost goes without saying that ladyhood was no small step for some of these girls. In fact, these girls presented many layers of social problems that a training school could ill afford to ignore. Twenty percent, MacNaughton reported, could neither read nor write, and 91 percent used tobacco. She described several girls as "snuff fiends." Ninety-one percent had served time in jail and 70 percent had worked in the mills, she claimed, in a condescending conflation of jailbirds and textile workers. And that was not all. "Ninety-seven per cent of the girls who come to us," she reported, "are diseased, varying from the mildest cases to the most virulent, capable of infecting a whole town." One girl, she explained, had begun work in a cotton mill at age ten. For two years she worked from 7 A.M. until 6 P.M. At twelve, she gave birth to a child by a married man. According to the girl, the man had money and "plugged the jury," who acquitted him. "Shunned and sneered at" in her hometown, the girl drifted into a "camp city where she made her living in the only way she saw open to her," MacNaughton explained. She was the kind of girl, "a victim of social circumstances," that Samarcand was pledged to save. "Our training," MacNaughton reported, "is to wipe out the past and make the girl feel that her future is hers to make of it what she will, that all has not been lost and that she may again take her place in Society."[4]

Yet MacNaughton never abandoned her dream to become headmistress of a respectable training school for young ladies. In 1921 she requested that Commissioner Johnson support a second women's reformatory so that Samarcand could operate "as a school for training young girls." A second reformatory, she postulated, would take the most difficult cases. In MacNaughton's way of thinking, any young woman there "capable of taking the training" and "really desiring to lead a different life" would begin at a correctional institution and later transfer to Samarcand Manor. Her ambition bred such temerity that she advised her own supervisor to redirect Samarcand's mission and lobby for a new reformatory elsewhere. "I feel that when you are speaking in the different communities," she instructed the chief of North Carolina's Department of Charities and Public Welfare, "that you can speak of us as a school and that there should be a reformatory just as there is in Pennsylvania and other states."[5]

If only the legislature shared her sense of urgency. By 1920, legislators were hardly forthcoming with funding, much less planning to build another reformatory. Indeed, Samarcand's financial legs were tottering. By early February

1920, lack of funds threatened to close the institution. Treasurer W. S. Blakeney reported that no money existed to pay the bills for November and December 1919 and that half of the 1920 state appropriation was "already exhausted." The only reason Blakeney had been able to pay the October 1919 bills was that the facility received a $10,000 advance from the Interdepartmental Social Hygiene Board to treat venereal disease. If the General Assembly did not soon pass a significant appropriation, he warned, Samarcand would be in "dire straits." He begged Commissioner R. F. Beasley to seek more assistance from the Interdepartmental Social Hygiene Board in Washington, D.C.[6]

The financial situation at Samarcand was in fact dire. In her own call for help, Agnes MacNaughton wrote her superiors of overcrowding, inadequate equipment, and ill-prepared staff. As of February 16, 1920, Samarcand housed 151 girls, at least two of whom were young mothers. According to MacNaughton, all but two girls "are infected with either syphilis or gonorrhea or both." No one had anticipated the medical expenses—the building, equipment, drug, and medical staff costs—needed for treatment in such circumstances. MacNaughton also recognized that actively infectious cases required segregation from other girls and that proper treatment often required a longer period of detention. The high incidence of infection, MacNaughton confessed, taxed the education of the girls, whose "senses are dulled by their physical condition." Furthermore, significant expense went toward purchasing additional kitchen equipment, bedding, and chairs beyond the original appropriation for thirty girls.[7]

Given the financial straits and the state legislature's hesitance to act, Commissioner Beasley negotiated nearly $60,000 from the U.S. Interdepartmental Social Hygiene Board. It was an agreement that, for Samarcand, one might liken to a pact with the devil. The sole purpose of the funding violated MacNaughton's prescription as a school for training young ladies and, instead, established the institution (at least in the eyes of the public) as a reformatory for fallen women. The decision for funding hung entirely on the degree to which the girls and the venereal disease they carried would present a "direct and certain menace to U.S. soldiers and sailors." To affirm the threat posed by the inmates, R. F. Beasley argued that seven thousand soldiers lived in the vicinity, and he later reported that this evidence "helped us a lot." When offered $20,000, Beasley pressed for more, again arguing the threat the girls posed to the armed forces. He asserted that "if we could get over the present emergency of the six months we could look to the state thereafter, but that unless we did get help one hundred inmates would have to be discharged."[8]

Within a week the Interdepartmental Social Hygiene Board had approved the funding "for the protection of the armed military and naval forces of the United States against venereal disease." Within six months a full $40,000 would be forthcoming from the federal government to protect U.S. soldiers from the disease-ridden girls at Samarcand Manor. Commissioner Beasley breathed a sigh of relief. "This help," he wrote in thanks, "will obviate a real disaster to our institution and to the whole cause of Anti-Venereal campaign in North Carolina." In fact, the $58,416.07 received from Washington in 1920 represented more than half of all the funds Samarcand received that year. It was marked on the financial statement as "Received from Fed for V.D."

FINANCIAL STATEMENT

Total Expenses Dec. 1918–Dec. 1919	$42,612.43
Total Expenses Dec. 1919–Dec. 1920	$68,004.95
Total	$110,617.38
Received from Fed for V.D.	$58,416.07 (+ Dr. Services)
Maintenance from State	$40,000
Total	$98,416.07
Borrowed from State	$12,201.31
Total	$110,617.38[9]

Despite its nefarious significance, the federal funds seemed to restore new rhythms of life at Samarcand Manor. That September, Agnes MacNaughton reported that the girls and the staff had adopted new routines to compensate for the overcrowded and inadequate conditions. MacNaughton had turned away some eighty girls, "all urgent cases," because of overcrowding. The residential halls were full and some girls slept on the floor. "Girls girls everywhere and hardly room to think!" MacNaughton remarked that month in a letter to State Public Welfare Commissioner Kate Burr Johnson. There were so many inmates that the staff had to develop activities to keep them all busy. Daily swimming and diving classes occupied the inmates and entertained visitors. One inmate, under the careful supervision and training of a teacher, joined the kitchen staff and eventually began to supervise the kitchen and prepare meals for two hundred people at a sitting. Use of an inmate in the kitchen was one answer to her labor shortage, but MacNaughton marveled at how quickly this girl learned her new kitchen skills. "This girl did not know one single thing in Domestic Science when she came to us," MacNaughton explained. MacNaughton even described the school routine as "lively and exciting as ever," though what she described were significant inconveniences that made it almost impossible to teach. The electrical plant, she explained had burst a month previous and "our only gasoline pump goes on a strike periodically." Obtaining clean water also was a problem. Girls and teachers carried water from "our well beloved spring" to provide for the full needs of the institution. As if to make matters worse, MacNaughton and the teachers had to contend with more distractions by young men following the girls from the spring. "Two sporty looking young men have been hanging on our grounds with the idea of becoming young Lochinvar and carrying off their lady love." Again the staff took to their revolvers and chased the young men from the premises. "Apparently we have dampened their ardor for the past few nights we have not been bothered."[10]

When the legislature failed to remedy these problems, MacNaughton turned to the press for help. T. E. Browne, a journalist for Raleigh's *News and Observer,* readily complied. In "Visit to Samarcand Manor," Browne promoted the worthwhile service of MacNaughton and her teachers in transforming criminals into healthy young girls. He described the modest campus in its salubrious rural environment, the hospital and staff, and the multi-use Manor building used interchangeably for administration, classes, kitchen, dining room, chapel, and sleeping rooms. He noted the "wonderful physical development of the girls." Once anemic, thin, and "with hard drawn expressions," the girls thrived under the positive routine, "regular hours and wholesome food." MacNaughton imposed a daily schedule that required girls not only to go to school, but also to work on the property. Divided into groups, girls worked on the farm, attended classes, and completed household duties. Those on farm duty cared for hogs, cows, and chickens, tended the crops, and dug ditches. As a result of the tight routine, "there is little time for mischief," Browne reported. The girls "do all the work of the institution, the cooking, making of their uniforms, caring for the rooms, etc., as well as the farm work and every girl must go to school at least three and one-half hours daily."[11]

DAILY SCHEDULE

6:00 a.m.—Rising bell
6:30–7:00—Breakfast
7:00–11:00—Farm group on farm
7:00–7:45—Household tasks
8:00–11:30—Classes
11:30–12:00—Recreation
12:00–12:30—Dinner
12:30–2:00—Rest or games
2:00–5:30—Farm work and classes
5:30–6:00—Recreation
6:00–6:30—Supper
6:30–7:00—Recreation—games
7:00–7:30—Chapel[12]

From the way Browne described it, the highly regimented schedule at Samarcand had a wholly transformative effect. These girls were once "victims of unfortunate circumstances, the product of depraved and unwholesome surroundings," and had never known the love or sympathy of a real parent. Many had gone from farm to factory at a young age. Browne described many as "subnormal mentally, but the majority bright and responsive without either the affectionate direction or restraint of a normal home." "Not only is it gratifying to see the wonderful physical change—never have I seen a healthier, more robust group," Browne remarked, "but it is an inspiration to observe the change of spirit, the new outlook upon life, that comes over these girls." With a genuine fondness for their teachers, they seemed happy and contented, "more like a crowd of rollicking children."[13]

Then Browne commenced a plea to the General Assembly. He congratulated the state for the fine institution and Mrs. MacNaughton for her "noble work." The legislators must "amply support it," he said, to provide for large numbers of girls. Once given a clean bill of health and some education, these girls would become useful members of society. "How much better it is for the State to spend liberally for such an institution," Browne concluded, "than to neglect these girls and let them ultimately become public charges either in insane asylums or criminal institutions!" Life at Samarcand Manor offered the requisite prescription for restoring healthy ladyhood to white girls led astray. The fresh air, household and farm chores, commitment to education (but not excessively so), chapel, and recreation offered girls the path to southern white womanhood. Girls properly educated and trained would become "useful" rather than an institutional drain. Agnes MacNaughton could not have asked for a more sympathetic treatment of her charges from the state's flagship newspaper.[14]

MacNaughton needed all the good press she could muster. Almost everyone, even girls' own parents, believed the girls at Samarcand were "ruined." One mother, the parent of "Mildred" from Parkton, North Carolina, agonized over her daughter's misfortune for landing in Samarcand. Mildred's parents owned a small mattress company. Her mother probably ran most of the business because her father was blind and disabled. She had had Mildred committed after a rape by a stranger. According to Mildred's mother, one night a man ("a beast of a man," according to the mother) who was boarding in a house across the street reached into a window beside Mildred's bed, pulled her out, hurried her away before she had fully awakened, and "committed an act that is unspeakable." Mildred's mother wrestled with where to place the blame for the attack. Apparently the rapist's culpability was not enough. Mildred's mother explained in a letter to Commissioner Johnson that she had tried not to blame her daughter. A southern lady was to protect her chastity, even at the expense of death. A mother's job was the same, to protect her daughter's purity. So Mildred's mother blamed herself. "If I had properly trained her, to play safe . . . she would not have to suffer as she does. Her sad misfortune would not have been." To avoid a scandal, Mildred's parents refused to "give her to marriage" and, instead, committed her to Samarcand though she had committed no serious crime. The world was unjust, Mildred's mother admitted. The rapist went free and the child was sent to a home, branded as a criminal. MacNaughton faced a steep hill to climb against both societal and parental perceptions of ruined daughters and criminal behavior.[15]

So MacNaughton recruited others to promote the idea that the reformatory could turn bad girls into southern ladies. An historian and archivist at the North Carolina Historical Commission, Fred A. Olds, penned a narrative for the *Orphans' Friend and Masonic Journal.* His goal was to promote the institution, and he constructed a sexualized narrative to that end. Olds presented the unreclaimed girl as a sexual menace to society. Samarcand, he said, reclaims the lost girl who is "perhaps poisonous as a snake," one who "might infect 500 men with that disease which crippled a formidable proportion of American young men called into their country's service." This disease, presumably syphilis, though Olds did not mention it by name, had cost the United States "over a million lost days of soldier service and twelve million dollars in money." Thus, Samaracand's significance lay in its cleansing effect. Many girls arrived infected, Olds argued, with a disease powerful enough to "poison and diminish American manhood, yes and womanhood too in a lesser degree." Samarcand offered the state the vastly important service of cleansing the female population of this dire threat to American manhood. "Think, North Carolinians," he urged, "of an infected child of thirteen years now made-over at this school."[16]

On the day of his visit, he observed a new girl arriving at the school. In a report to his readers, he explained the process of her reclamation. Up she came, he said, in an automobile driven by a welfare officer from a county where she "had been caught." Olds neglected to mention her crime, but he certainly was fascinated by her appearance. She was barely seventeen, he noted, from a small town, and "of her family the less said the better." She was not just a wayward girl, but a teenage reprobate, a sexual seductress. "Slim, with a near-velvet jacket, a skirt only a few inches below her knees, and silk stockings liberally displayed and shoes with heels nearly three inches high, she looked as if she scorned all the world." She stood in the doorway of the dining-room, bright with electric lights. "I won't stay here," she informed the county welfare officer, but he told her "Good-bye" anyway. At Samarcand, the able medical staff treated her two infections, "both of which are slayers of man and woman-power." Her "rosy-cheeked" peers taught her the daily routine so that "in the coming days and months she was to be made clean too instead of impure, and helpful instead of poisonous." Her hair was washed and she donned the light blue costume worn by probation girls and new arrivals.[17]

The pastoral milieu of "mile upon mile of peach trees" must have helped improve this barely seventeen girl. Yet to Olds, the process of reclamation was far more complicated than simply placing a girl in a healthy environment. The staff at Samarcand also had to restore her lost virtue. A new girl, he noted, was placed on probation, and she would have to wear the requisite blue probationary uniform. After three months she could earn credit toward "honor girl" status by her good behavior. Honor girls participated in student government and wore light brown khaki instead of blue. They had to recite three pledges: "one loyalty to the school, the second to do honorable things and the third to make at least one person happy each day." The very best, most virtuous girls helped with the ultimate challenge—rebuffing the ubiquitous and uninvited male visitors who prowled around the place. When "one gang" parked their automobile in front of the barn, cut the telephone wires, and lit a fire, some of the girls eagerly helped to chase off the "Raiders." Were the girls' intentions virtuous? Who really knew, but Olds presumed so. "The student government girls' behavior was wonderful," Olds remarked for the occasion. Yet he was aware that not all was right at the institution. Samarcand still suffered its share of naughty girls, as evidenced by periodic runaways, though no girl, he noted, had escaped without recapture. Instilling virtue was not without some failure.[18]

No girl, it appeared, was entirely lost. At least, that is what the administration led people to believe about the delinquent white girls in its charge. In a 1920 speech in Goldsboro, North Carolina, MacNaughton read an extract of a text supposedly written by one of the girls to demonstrate that her chief aspiration was to achieve ladyhood. "The whole philosophy of our girls seems to be summed up in this extract from one of their letters: 'I have done dirty but that is no reasin I can not make a lady out off myself every lady hase had failures and have had mine but it will come out all right some sweet day and I know I can bee sum lady and I am a goying to.'" One journalist applauded Samarcand for returning the state's financial investment "manifold in the salvation of fallen womanhood." Samarcand Manor seemed a success. This journalist concluded that "the State should now prepare itself to rescue all scarlet women" who threaten to become charges on the state.[19]

Even the deplorable conditions built character. Lack of a laundry and an unreliable water system forced girls to wash clothes at the lake. Olds might well have described the setting as sublime. It was a picturesque scene where girls returned to nature to learn the meaning of hard work. "At a point where the stream as clear as a dew-drop rushes from the lake are their wash-pots, under the graceful long-leafed pines, and a little beyond is the place where the clothes are spread to dry." The girls worked hard, looked healthy, and were charming in their vitality. "No namby-pamby girls are these," he said. Agnes MacNaughton found the circumstances of laundry at the lake just a bit less inspiring. While the girls enthusiastically volunteered for laundry duty and a chance for a quick swim, losing the water supply was a major inconvenience as it also entailed toting water from the spring and up the hill.[20]

All that winter and into the next spring girls washed their clothes at the lake. They were still at it in July 1920 when Kate Johnson sent an inspector to investigate complaints by former teachers that had spread as far north as New York City and as far west as St. Louis. This word had come from an Asheville reformatory director who had heard secondhand from a Missouri friend that conditions at Samarcand were "bad—that the girls there were not taught anything along school lines and little of value along the other lines." Alarmed by this news, Johnson promptly hired Grace A. Reeder, assistant superintendent of the New York Orphanage at Hastings-on-Hudson, to staunch the rumors by conducting an investigation at Samarcand. Reeder possessed all of the qualifications in the profession; she was a graduate of Smith College and Columbia University, she was experienced in institutional management and administration, and she had conducted these sorts of investigations before. As an outside source, Reeder offered an objective critique that people in the field could trust, and a favorable report from her would quelch the gossip that had begun to spread far and wide.[21]

Disguised as MacNaughton's "guest," Reeder conducted her investigation during a three-day period in July 1921. She quickly learned that Samarcand was not at all an educational institution, but a reformatory with a mediocre industrial training program. The school's plan was to offer each girl three months of training in eight areas of domestic work, including "plain sewing," dressmaking, "domestic art" such as needlework, farming, milking, feeding pigs and chickens, laundry, and cooking and pantry work. Reeder noted that recordkeeping of girls' training was anecdotal and depended largely on the memory of the supervisors and the girls themselves. Furthermore, many girls did not get kitchen experience because inmates with venereal disease were prohibited from any kind of kitchen work or food preparation. In cooking there was no "theoretical training" and no classes, though Reeder had few expectations "considering the type of girl in the school." Her only expectation was that the girls learn how "to cook plain food well," and she felt sure they were getting that training. Training in dishwashing and pantry work, Reeder said, "leaves a good deal to be desired" because of the lack of equipment. Girls washed dishes in pails on the wooden tables in the dining room, and carelessly, she added. "The dishes are not rinsed and water is splashed around on the tables." Reeder reserved some praise in other domestic pursuits. Girls still washed clothes with cold lake water. Their work was good, she added, and "the girls are learning to do laundry work and I had proof of it in a beautifully laundered waist which one of the girls did for me." Likewise, girls showed some promise in dressmaking, though the institution employed only a former cook in that line of work. Girls with "aptitude" assisted in the hospital, a new millinery class, or in assisting with the "morning athletic drill." But all the girls learned to make beds, darn stockings, scrub floors, clean, and perform the daily athletic drill.[22]

Yet for all that, Reeder criticized the girls' lack of womanly behavior. Although focused on teachers and training, Reeder's report also offers a raw perspective on the girls' experiences that none of the promotional newspaper articles or MacNaughton's own reports could provide. Each night at dinner, she sat opposite the "discipline group." Girls sat at long wooden tables and, knowing that the same tables were used for washing, made little effort to keep their food from spilling off their plates. Their table manners apparently left much to be desired. Reeder observed girls "eating with both elbows on the table and heaps bent over their plate." Many ate everything, including vegetables, with a spoon. Probably due to the inconvenience of washing at the lake, the staff provided no tablecloths or napkins to encourage more refined behavior. Though a small matter, she acknowledged, "this will be the only chance the girl will have to learn some of the refinements of life." Reeder recommended paper napkins at the evening meal and discussions with girls on proper table manners. One other area of refinement also required immediate attention, that of dress. Girls wore bloomers all day and attended meals in the same suit in which they worked on the farm or in the barn. For many, Reeder reported, slovenliness had been the general rule at home. Thus, Samarcand should teach them to comb their hair and to put on a clean dress once a day. Better habits in dress will raise girls' standards in personal appearance when they leave. "Beside the matter of neatness," Reeder observed, "the change to dresses once a day will help to overcome the tendency to roughness and mannishness which the wearing or [sic] bloomers nearly always brings out in girls who have not had the proper home training." The staff, she concluded, should make training in womanliness a priority.[23]

In addition to the prerequisite ladylike behavior, administrators encouraged a degree of self-government or citizenship training. Reeder reserved high praise for the student government or "honor" system that MacNaughton had established at the school. At the time of Reeder's visit, the girls and staff referred to this association as the "Democracy." Girls earned eligibility as "honor girls" after three months' residence and upon demonstrating good behavior and loyalty to the school. Honor girls voted one another into membership at weekly meetings and elected a student body president under the supervision of a staff member. It had been reported to Reeder that the "Democracy" elected sixteen girls to a council that governed many of the school's activities and supervised girls at recreation, and, she said, they were "proud to show themselves worthy of the trust placed in them by Miss MacNaughton." Perhaps this demonstration of citizenship left its greatest impression upon the inspector. In her final evaluation, the training in citizenship appears to have muted all of her other concerns about Samarcand. Indeed, her report raises numerous questions about training and education at the institution. But in the end, she concluded that "with the exception of the lack of training in table manners and the mater [*sic*] of neatness in dress, it would seem to me that the girls are getting excellent training even with the present equipment and that North Carolina may well be proud of the work being done at Samarcand Manor."[24] Reeder's final positive assessment contrasts with her less effusive daily reports, but suffice it to say, her report praised elements of Samarcand's industrial training and honor system while criticizing the academics and table manners at the institution.

Grace Reeder's investigations and conclusions underscored the inflexibilities, inconsistencies, and unrealistic expectations in the requisites of southern white womanhood. According to the ideal, southern white ladies were intrinsically mannerly. Reeder had discovered Samarcand girls who lacked manners, including the skills of using utensils at the dinner table. Southern white ladies were pure and chaste. Reeder commented on the prevalence of venereal diseases that stained Samarcand Manor girls' virtue. Southern white ladies were submissive and obedient. Reeder praised Samarcand Manor for its honor system, the "Democracy," that encouraged leadership and citizenship among girls. This last component, the honor system, was the most inconsistent in the manner of training southern ladies. In a state that had refused women's suffrage, the honor system created by Agnes MacNaughton encouraged leadership and civic participation, but only among the good girls who had shown excellence in purity, submissiveness, and proper manners. Bad girls who had not met the standards of ladyhood were not trusted with the reins of civic participation.

In 1920, women's suffrage was the law of the land. In the South, where Jim Crow laws still restricted African American women's political participation, white girls could expect to vote upon reaching the age of twenty-one. MacNaughton leapt at every opportunity to promote her reformatory as a model of training in citizenship to the press. In 1922, she invited yet another journalist, Mrs. C. A. Loop, to the campus for a visit. Mrs. Loop duly reported on the usual topics of interest, including the status of construction, the lovely grounds, and the nutritious food. Yet, as is apparent from the title of her article, "A Home and Not a Prison: A School Where the Honor System Rules," the student government association was her chief subject. According to Loop, the student government system structured discipline and punishment at Samarcand Manor. Thanks to its very careful implementation, Samarcand operated without bars or padlocks and girls "could come and go as they pleased." In each dormitory lived at least two teachers and at least two council members. Any matter of discipline came before the council, who decided upon punishment, subject to the superintendent's authority. Even runaways were subject to this governing body. Loop noted that the council knew of any escape within twenty minutes. She did not mention that by the nature of the very act, news of a runaway probably spread like wildfire among the general student body anyway. Instead, she argued that the "Democracy" was a humane and fair system. "Although there is gentleness in all dealings," she noted, "there is behind it a firm determination that all rules shall be respected which of itself demands respect from the girls."[25]

Yet Mrs. Loop's laudatory praise reveals certain truths about the self-government association. MacNaughton had developed an illusion, a scheme in which good girls—those who behaved and did not attempt escape—enjoyed privileges that bad girls could not. Meanwhile, MacNaughton retained tight control over discipline. She maintained ultimate authority over punishment and selected the sixteen council members herself. Good girls lived in "honor cottages" and bad girls lived in the "receiving dorm," also known as Chamberlain Hall. Good girls enjoyed picnics in the woods, wading parties, hikes, attending church and movies, and planned weekend camping parties in the summer. Bad girls participated in none of these. "We find," MacNaughton observed, "that Student Government is the best incentive for a girl to make good."[26]

A good girl knew the rules. She recited them daily. Of course, the rules were hard to miss, as staff instilled them into the inmates at every turn. Girls and their teachers wrote songs, mottoes, and creeds for their residence halls to encourage comradery and loyalty to the school. Mottoes encouraged orderliness, kindness, truthfulness, purity, and loyalty. One poem, "Student Government Don'ts," enshrined the rules in verse:

Don't laugh while in preaching,
Don't cry at a ball;
Keep silence at rest hour,
Don't walk in the hall.

Don't be an old slacker,
Don't roll your hose,
Be courteious [*sic*] at all times,
Don't argue who goes.

Don't be behind time,
Don't curl your hair;
Where you're not wanted—
Don't even go there.

Don't ever use powder,
Don't rouge your face—
Your nose may be shining,
Don't think it a disgrace.[27]

What happened to a girl when she broke the rules? When summoned to a council meeting, a girl met a shock. The summons came from a residence hall proctor and she was expected to appear at once. At the council meeting, she came face to face with the school president and council members from every hall on the hill. The president assumed a serious air and established the tone of the meeting. The stakes were high. Every girl knew that a disciplinary action meant more than a Saturday of lost fun. Girls on the offender's hall would also lose at least one point for her carelessness. Oh now how easy it would seem to refrain from talking in class. The council presented the charges and heard the case. Then the president asked the offender to leave the room for five minutes. My, how long those five minutes were. The proctor appeared for her once more. The president announced the council's decree: "you shall join the Saturday discipline group for the next two Saturdays." The punishment might vary from week to week. While good girls engaged in a candy pull or a hike in the woods, bad girls stayed behind to clean the barn.[28]

The staff encouraged good girls to turn against the bad girls. Really bad girls suffered confinement and endured whippings by staff or other girls. In 1926, May McCubbins's parents visited their daughter, who shocked them with a horror story about the discipline she received at Samaracand Manor. Sixteen-year-old May from Valle Crucis had arrived at Samarcand with venereal disease. Agnes MacNaughton described her as a psychopathic case, chiefly because she fought constantly with the other girls. According to May, the fights occurred because the honor cottage girls harassed her about her venereal disease infection. May slapped girls in class and frequently ended games on the athletic field by breaking into fights with other girls. Upon her parents' visit, May complained that she had served four months in a confinement room and endured five separate whippings while at Samarcand, twice by the other girls at the consent of a staff member. On one occasion, thirteen girls whipped her with their hands and a hair brush. On another occasion, May reported, she was whipped with "some kind of a plaited reed or lash." She showed her parents bruises on her hip, her back, and marks on her legs. She later complained that staff had whipped her for confiding to her parents.[29]

The McCubbinses demanded an investigation. Agnes MacNaughton consented and finally determined that most of May's complaints were false. Girls testified that May pinched herself to make the marks. May admitted that she fought and that she refused to "make good at Samarcand" because she wanted to leave. MacNaughton concluded that May had overstated the confinement and whippings. Yet others also had complained of abuse. Girls, with or without consent of the staff, took discipline into their own hands by harassing girls who did not conform. In March 1929, another parent, a mother of a girl from Parkton, North Carolina, described the form of justice that girls meted out to one another. This parent also said that girls were punished for complaining to parents. Her own daughter had told her that girls were forced to lie face down while all the girls in the cottage took turns beating her. According to this parent, "if a girl refused to do her share of this licking, she must also take a severe licking for not licking the girl or girls being punished." MacNaughton expressed "great regret and mortification" that girls whipped each other.[30]

Bad girls suffered confinement and exclusion from activities of all sorts. What was the most exciting event at Samarcand that a bad girl could miss? Annual field day. In 1922, one observer explained that events began at 9:30 A.M., when all the "honor cottage" girls lined upon the field in their light tan bloomer suits. Each girl wore a number on her collar for the track event, and a band on her arm denoting her honor cottage. Girls competed first in the running broad jump. They ran the fifty-yard dash, leaped at hurdles, and exercised their skills at the baseball and basketball throw. The vaulting event permitted competitors three attempts. Frenzied shouts accompanied the most exciting morning event, the hall relay race. Girls participated on teams organized by cottage. What yells there were! Participants and spectators cheered exuberantly as Carroll Hall finished first. A hush fell over the field as girls lined up for the athletic drills. Some drills were conducted by rank, in military fashion. In one impressive display, girls filed into the center of the field to form a huge wheel which completed one revolution. In a most exciting denouement, everyone saluted as the youngest inmate, a five-year-old girl (probably committed to Samarcand as a wayward orphan), carried an American flag onto the field. A lightning storm and a big clap of thunder interrupted the final baseball game. "Everyone's attention had been given so completedly [sic] to the game that no one saw the darkening sky. In a minute the deluge came and a minute after that, the Baseball Diamond looked like a small size edition of the deserted village." After the shower, everyone enjoyed lunch in the pavilion, live music, and a play. Swimming and diving contests followed. "There is always something about a race," one observer remarked about the swimming competition, "that appeals to everyone." Except that on field day the bad girls were not invited.[31]

Girls relished the rivalry and sport of field day. Running, jumping, dashing like boys, girls in bloomers competed with ease. The bloomer suit, or bifurcated skirt, was a pantaloon that extended below the knee. It liberated girls and women from the restrictive corsets and other garments that had confined women of earlier generations into sedentary lifestyles. It appeared in the mid-nineteenth century simultaneous with the first public discussions of women's rights and suffrage. The bloomer was not without controversy. In the nineteenth century, women who wore the garment in public were ostracized, hissed, and hooted, even if accompanied by their husband or a brother. Even early twentieth-century observers such as Grace A. Reeder found the costume mannish and unwomanly. Perhaps so. Yet feminists endorsed the bloomer suit as a significant contribution to women's physical education.[32]

The bloomer suit gained popularity thanks to feminist arguments about women's and girls' education. Feminists argued that physical vigor promoted mental acuity and good health. They made these arguments in response to decades of unyielding claims by medical professionals who said that physical activity and mental exertion of any sort inhibited women's sexual development. Male physicians such as Henry Maudsley argued that proper sexual development required girls and boys to preserve their "vital energy." In sum, boys expended their vital force by ejaculation, and girls did so by physical and mental exercise. Feminists labored tirelessly to prove that physical exertion instead promoted healthy and strong mental constitution in women. Women's education, both intellectually and physically, were thus ideologically intertwined. One scholar has also noted the other side of the coin of the promotion of women's health. The controversy over women's health sparked an almost obsessive concern with girls' individual bodies, as professionals on both sides of the debate scrutinized girls for any weakness or deformity.[33]

Annual field day, then, carried a double meaning. It at once rewarded student government girls for good behavior and served as a feminist statement toward girls' and women's physical development. The event celebrated women's physical activity and feminist dress and represented an important and exciting outlet for honor girls to express themselves. Yet, administrators used annual field day as a message to the outside world. To anyone watching, Agnes MacNaughton had appeared to transform the naughtiest white girls in the state into robust, rosy-cheeked, modern young women. The truly naughty girls, those who refused to conform to the Samarcand model, went unseen. The two-tiered governing structure thus presented a picture of success. To the press and the public, the reformatory reclaimed fallen women and restored them into southern ladies. In reality, it strictly regimented those girls who conformed to the rules of behavior and cordoned off those who did not. Of course, there was some fluidity. Girls on probation could earn a transfer into an honor cottage. Honor girls could commit enough infractions to send them into confinement at Chamberlain Hall.

One evening in May 1922, the honor girls proved their mettle. It had been a hot and sultry morning. By rest hour, according to one account, "everyone was wilting." At 3:30 the sky turned glossy yellow and black clouds loomed on the horizon. "By 5:00, it was dark as night." Thunder rumbled in the distance and lightning flashed. "And then it seemed as tho the sky had been forced wide open and banged shut." Everyone sat silent, until someone whispered, "That must have struck somewhere near here." In fact it had. Lightning had struck a transformer, and fire sped along the wires leading into the Manor. Within moments the roof was ablaze. According to one account, girls from the honor cottages leapt into action. Every girl turned out with buckets and pots, but it was too late. Girls, both the foolish and the brave, dashed into the building's first floor to rescue what they could. Papers, furniture, the safe, books, and other goods made it out. That was all they could save. Before long, the Manor's west end was destroyed. The mass of flames seethed and hissed. Windows crashed, cartridges in the revolvers exploded. But the most fearsome sound was the wind. Fearing the fire might jump the trees and spark something else, the girls turned their buckets of water and sand on the Manor and the hospital. With a crash, the Manor's frame collapsed. When the Pinehurst Fire Department arrived, there was nothing left to save. Girls returned to their cottages. The Manor had become a memory. No account of the 1922 fire mentions the bad girls that night. As the story goes, it was the good girls who had made the save.[34]

FIGURE 1. Women's club reformers advocated the establishment of the State Home and Industrial School for Girls to train "problem girls" into young southern ladies. The reformatory did not admit girls of color. These girls are "graduating" from school, yet remain inmates until paroled. No date. Records of Samarcand Manor, Division of Adult Correction and Juvenile Justice, Department of Public Safety, Samarcand Manor School, Eagle Springs, North Carolina.

FIGURE 2. At Samarcand Manor honor girls wore khaki dresses and bad girls wore blue dresses. The blue dress designated a girl under disciplinary supervision and living in a discipline dorm, such as Chamberlain Hall. Samarcand Manor, Division of Adult Correction and Juvenile Justice, Department of Public Safety, Samarcand Manor School, Eagle Springs, North Carolina.

FIGURE 3. Staff produced portraits of these girls in the early 1920s. They were not inmates at the time of the trial. However, the pictures serve as representations of the girls who were committed to Samarcand. Records of Samarcand Manor, Division of Adult Correction and Juvenile Justice, Department of Public Safety, Samarcand Manor School, Eagle Springs, North Carolina.

FIGURE 4. At Samarcand Manor, promotional literature portrayed the reformatory as offering the essential training for white ladyhood, including proper hygiene and religious instruction. Circa 1922. Samarcand Manor School Records, Mars 5899, Department of Public Safety, State Archives of North Carolina, Raleigh, North Carolina.

FIGURE 5. Samarcand Manor, 1921. Fire destroyed this Samarcand Manor administration building in 1922. Staff records attributed the cause as lightning. Samarcand Manor School Records, Mars 5899, Department of Public Safety, State Archives of North Carolina, Raleigh, North Carolina.

FIGURE 6. Agnes MacNaughton served as the first superintendent of Samarcand Manor, 1919–33. Her administration adopted the merit, honor, and corporal punishment system. She quietly resigned after the 1931 arson trial investigation. Samarcand Manor School Records, Mars 5899, Department of Public Safety, State Archives of North Carolina, Raleigh, North Carolina.

FIGURE 7. In 1922, Agnes MacNaughton reported that girls volunteered to do laundry at the lake for a chance at a quick swim. Samarcand Manor lacked a laundry and had an unreliable water system. Records of Samarcand Manor, Division of Adult Correction and Juvenile Justice, Department of Public Safety, Samarcand Manor School, Eagle Springs, North Carolina.

FIGURE 8. This 1930 photograph, taken the year before the arson trial, illustrates only about one hundred inmates when Samarcand's population averaged about two hundred girls per year. The honor girls are filing out of the church, youngest first. The girl second in line on the right is dressed in tan, likely an indicator that she is a new admission and serving a requisite term of probation. Records of Samarcand Manor, Division of Adult Correction and Juvenile Justice, Department of Public Safety, Samarcand Manor School, Eagle Springs, North Carolina.

FIGURE 9. Maypole Dance, 1921. At Samarcand Manor, the administration permitted good girls to participate in May Day events. Throughout the 1920s the maypole dance was a popular activity at the reformatory. Samarcand Manor School Records, Mars 5899, Department of Public Safety, State Archives of North Carolina, Raleigh, North Carolina.

FIGURE 10. This 1926 photograph is titled "Play Time" and was used as promotional material for the reformatory. The photo indicates that some of the inmates were under ten years old. Records of Samarcand Manor, Division of Adult Correction and Juvenile Justice, Department of Public Safety, Samarcand Manor School, Eagle Springs, North Carolina.

FIGURE 11. On field day, girls participated in sport activities including track, baseball, and fencing. These girls are wearing pantaloons, a costume urged by feminists as necessary to a woman's participation in sport for good health. Records of Samarcand Manor, Division of Adult Correction and Juvenile Justice, Department of Public Safety, Samarcand Manor School, Eagle Springs, North Carolina.

FIGURE 12. This 1942 May Day photo shows that girls continued to engage in theatrical productions and sport activities, as had inmates twenty years earlier. Records of Samarcand Manor, Division of Adult Correction and Juvenile Justice, Department of Public Safety, Samarcand Manor School, Eagle Springs, North Carolina.

FIGURE 13. The staff encouraged good girls to engage in theatrical productions and allowed them to attend movies on special trips to town. On these outings girls would have seen their favorite female film stars on the big screen. Circa early 1940s. Records of Samarcand Manor, Division of Adult Correction and Juvenile Justice, Department of Public Safety, Samarcand Manor School, Eagle Springs, North Carolina.

FIGURE 14. Kate Burr Johnson, about 1930. In 1921, Governor Thomas Bickett appointed Kate Burr Johnson, a former president of the North Carolina Federation of Women's Clubs, as the first female state commissioner of public welfare. She served until 1930. Samarcand Manor School Records, Mars 5899, Department of Public Safety, State Archives of North Carolina, Raleigh, North Carolina.

4

Suddenly Proclaimed Unfit

THE EUGENICS AGENDA OF KATE BURR JOHNSON

"KATE HAS A HEAD FULL OF BRAINS and she knows how to use them," observed Governor O. Max Gardner about the esteemed Kate Burr Johnson. It was a deserved compliment. Johnson was the first female state commissioner of public welfare in the nation. Governor Thomas Bickett had appointed her in 1921 when she was forty. Already she had accumulated a long record in civic service and public welfare. Her resume included being president of the North Carolina Federation of Women's Clubs, director of the Board of Charities and Public Welfare, and a member of the North Carolina Equal Suffrage League. In her nine-year tenure as commissioner, Johnson would mastermind the state's identification, isolation, and treatment of the feebleminded and serve as the architect of the state's notorious sterilization protocol.[1]

Johnson fancied herself a social scientist. As a former women's club reformer, she possessed an uncommon understanding for the poor and downtrodden, but as a mental hygienist she also expressed an unapologetic condemnation for the most disadvantaged individuals in society. She embraced both the disdainful language of the eugenicist and the optimistic view of the progressive reformer. As a eugenicist, she argued that the cause of poverty and social malaise originated in undesirable hereditary characteristics that passed from one generation to the next. Yet as a progressive reformer, she betrayed the eugenicists by arguing that environmental conditions, not innate breeding, caused social problems and individual maladjustments. How did she reconcile these two apparently contradictory ideologies? To rephrase the question, how could she profess both nature and nurture? The answer lies in her racial beliefs. This chapter argues that mental hygiene and eugenics influenced Kate Burr Johnson's understanding of southern white womanhood and shaped her vision of progressive reform.

She argued that although southern white girls were preconditioned for lady-hood, environmental dangers, specifically immorality in the form of prostitu-tion and illicit interracial sexual relations, threatened the genetic codes that defined whiteness. Behind Johnson's Anglo-Saxon face was a mind that burned to preserve the white race, a responsibility that rested in the control of white girls' sexuality through state enforcement and institutionalization of those pro-claimed "unfit." She argued a Lamarckian approach to eugenics that embraced whiteness as genetically superior but also emphasized that a bad environment, especially when experienced by white women, produced irreparable genetic damage and resulted in mental defectiveness in offspring that threatened the white race. As commissioner of public welfare, it was not only her duty to reform the poor, dependent, and neglected, but to save civilization from the increase of the most unfit among them.

Johnson was born and educated in North Carolina. As a child, the Appa-lachian Mountains near Morganton provided her a playground. As a young adult, the metropolitan setting of Charlotte and Queens College inspired her intellect. At the impressionable age of twenty-four, when one begins to collect experiences that shape one's adult intellect and identity for a lifetime, she at-tended graduate school in the pastoral setting of Chapel Hill. Many observers noted her refinement, cultivation, and charm. Yet what truly impressed them most about Johnson was the mark she made on the social work profession and the direction in which she sent public welfare in her nine-year tenure as commissioner. One journalist cited her as "the spirit of progress" in North Carolina. Described as a "master builder," her achievement lay "not in stone or brick but in men and women, molding the warped and twisted, the derelict and forsaken, into self-respecting human beings who could bring strength in-stead of weakness to our society." She was a "good administrator," a "coura-geous leader," an "energetic worker," and a woman of unparalleled sympathy and compassion.[2]

She embraced the social philosophy in progressive reform. Progressive re-formers drew from the ranks of college-educated men and women who, com-mitted to civic virtue, applied the study of social science to an understanding of urban problems. Many reformers and settlement house workers had read the standard text, *Young Working Girls* (1913), by Robert A. Woods and Albert J. Kennedy. These two men argued that rapid industrialization and rapid urban-ization had disorganized society and the family so that adolescent girls had no moral compass to guide their behavior. According to this view, working-class girls lived only for individual pleasure and were highly susceptible to immoral stimuli. They explicitly critiqued poor parents, working-class neighborhoods, and big business for failing to supervise daughters in an urban society characterized by unregulated workplaces, dance halls, and other public amusements that threatened girls' virtue. Progressive reformers labored to restore this supervision with "protective" policies that restricted alcohol sales, regulated dance halls, and prohibited prostitution. The YWCA and other clubs fostered acceptable forms of entertainment and recreation. Above all, reformers contended that government and civic regulation would provide the supervision and protection that working-class girls so badly needed.[3]

Like the progressive reformers, Johnson promoted the idea that the urban environment contributed to the girl problem. In her second biennial report, covering the years 1922–1924, she argued for greater improvements to the state institutions and agencies that cared for the "neglected, delinquent and defective elements of North Carolina's population." These dependents came to the institutions as the result of "unfortunate social conditions," she said. In her report she recognized numerous causes of delinquency and dependence, all environmental. "Unsatisfactory economic conditions, poor educational facilities, inadequate programs of public health, public welfare and social and mental hygiene, these are only a few of the factors that go into the making of dependents, delinquents and defectives that institutions must care for."[4]

However, it is at this point that Johnson's argument departs from the more mainstream branch of progressive reform. Johnson claimed to be a mental hygienist, a member of a new, small (but influential) branch of progressive reformers. Mental hygienists accepted more permissive standards of sexual expression among adolescent girls. Unlike clubwomen, they rejected notions of female sexlessness and taught that the sexual impulse in girls was natural. In fact, some argued that girls required proper sex education to reinforce the idea of sex as a beautiful and spiritual force within marriage. They chiefly argued that delinquency and mental illness were related; both grew from mental deficiencies that originated in poor supervision or repressive behavior by parents. Mental hygienists typically employed the language of psychiatrists and psychologists of the 1920s. Girls who exhibited antisocial behavior were "maladjusted" and, with proper professional diagnosis, their path toward serious delinquency could be averted by counseling and other constructive "adjustments."[5]

Yet there lurked a nefarious underbelly to the mental hygienists' seemingly progressive attitudes about sexuality. While Johnson and other mental hygienists defended as healthy a normal range of sexual activities in girls and women, they sounded shrill alarms against sexual deviance, particularly in the nation's population of the feebleminded and delinquent. In her 1922 biennial report as North Carolina's commissioner of public welfare, Johnson warned that the dependent, delinquent, and feebleminded are "drags upon their communities, clogging the wheels of economic progress." She used other derogatory language to refer to this underclass; "parasites," "inferior," and a "social blight." She sounded dire warnings about their profound threat to society. In the wake of their loose and licentious sexual activity, defectives imposed upon society all of the evils of prostitution, venereal disease, and illegitimacy. "The children of feebleminded parents will be feebleminded," and we pay the costs, she warned. Current institutions, she argued, were inadequate to address "the pitiful hordes of the defective, i.e. the feebleminded, the insane, the epileptic, the deaf, dumb and blind." Upon assuming office, Johnson's first act as commissioner was to segregate the defective and thus prevent "promiscuous breeding." She reorganized the four bureaus of the State Board of Charities and Public Welfare (County Organization, Child Welfare, Institutional Supervision, and Promotion and Education) and vested each with certain supervisory powers over this dangerous population. Then, in September 1921, she added a fifth bureau, the Bureau of Mental Health and Hygiene, to report on the medical status and testing of North Carolina's defective population.[6]

Johnson followed a scientific practice known as eugenics that fathered the ideas behind psychiatric observation and intelligence testing. Eugenicists chiefly agreed that mental ability was hereditary and that society should strive to decrease the number of feebleminded, insane, handicapped, and other defectives at large in society. An English scientist and cousin of Charles Darwin, Francis Galton, had adopted the name "eugenics" from a Greek root meaning "good in birth" or "noble in heredity." Eugenics denoted a science for improving the human population by giving "more suitable races" a better chance of surviving and reproducing than those of lesser qualities. Johnson also embraced the notion of the inheritability of mental defectiveness. The root of all social problems, she asserted, including crime, immorality, and dependency, stemmed not from sin but from sickness or brain deficiencies. To allow deficiencies to circulate unchecked in society threatened the whole human stock.

When a Durham orphanage released nine allegedly "feeble-minded" children back into the community, Johnson protested. To allow a feebleminded child to resume a life unsupervised in society at large "is decidedly dangerous to the welfare of the community," she argued. Institutionalization offered a humane solution, she asserted. But the state could do more. Johnson meant to be at the head of any movement to control the worst of the human stock. "Segregation is imperative," she argued in her biennial report, "but, in addition, there should be humane measures that will prevent the reproduction of the patently feebleminded."[7]

Kate Burr Johnson's eugenics agenda is important to a study of the girls at Samarcand. Her understanding of eugenics fundamentally shaped her method of restoring white girls into southern ladies. A powerful woman in North Carolina's bureaucracy, she applied her views of white women and eugenics generally in her work with the poor, and specifically in the policies she fostered to control the reproduction of poor white women and girls. Her ideology began in the language of the progressive reformer. Delinquency, she argued, developed among white girls chiefly from poor and deleterious environmental conditions, such as poverty. With state intervention, white girls might be saved from poverty and juvenile delinquency. At this point her thought process shifted from the language of the reformer to the language of the eugenicist. Girls who persisted in bad environmental conditions in effect suffered genetic mutations that no longer resembled the purity of the white race. These genetic mutations, characterized by delinquent behavior, feeblemindedness, and immorality (defined as being sexually promiscuous, especially across the color line), effectively poisoned their whiteness. These traits eventually infected their genetic makeup and passed to their offspring. In her illustrious career as a social worker, Johnson sounded the alarm: poverty had created an urgent problem in the spread of the immoral feebleminded. Redeem these girls now or face the disastrous results of genetic mutations that threatened to infect the pure white stock. Wayward southern white women, Johnson concluded, could be redeemed as southern ladies as long as their good genes remained intact. Once the genetic mutations took place, they passed the point of no return.

Johnson's argument on the interplay of culture and biology reflected a movement particular to southern intellectuals in the era of eugenics. Her emphasis on the environment and its alleged role in genetic mutation echoes the work of southern social scientist William Fielding Ogburn. Ogburn, a sociology professor at Columbia University who was born and educated in Georgia, argued that "eugenics cannot be fairly estimated without a generous consideration of the cultural factor." Ogburn is best known for his "cultural lag" theory, which posits that society has trouble adapting to technological innovations due to social problems and maladjustment. His emphasis on culture also has important racial implications. "Man as we see him and know him," Ogburn wrote, "is always a product of two factors, heredity and environment." Behavior in culture is the product of social heritage (the habits, traditions, and material culture of a society) and biology. He criticized eugenicists who argued that genetic mutation accounted for most social problems. Ogburn argued that culture, not biology, accounts for behavioral differences within the white race. He maintained that where race remained constant, culture explained certain behavioral differences within the race. Northern Europeans, he said, bore such striking physical resemblance to the "old American stock" that perceptible behavioral differences are merely cultural and not biological.[8]

It was a subversive argument in the world of eugenics. Ogburn, Johnson, and other southern eugenicists challenged the national eugenics movement. In the national context, eugenicists did not always stress genetic mutation theories in racial terms. If poverty was genetic, they figured, then poor Americans, regardless of ethnicity, simply had inherited bad genes. By this logic, "poor whites" were of enfeebled stock. But Ogburn and Johnson wished to account for the genetic purity of the white race. Biology ensured the shared purity of white Americans and northern Europeans. They accounted for any behavioral differences as cultural; culture explained most variances among whites, rich or poor, on either side of the Atlantic. Poverty among whites thus was situational unless it persisted for generations, where it enfeebled the race.

Although she called herself a mental hygienist, Johnson's eugenic agenda rested squarely in an ideology of race. She argued that segregation of mental defectives and African Americans was critical to white purity. The commissioner was a very strong proponent of Jim Crow segregation because she feared that African Americans' inferior genetic characteristics might taint the white race. In her role as commissioner, Johnson viewed herself as being on the front line in the battle to preserve white civilization. Monitoring girls, particularly those most inclined toward sex offenses, remained one of her most important duties. Even delinquent boys did not require the same attention. The consequences of girls' sex offenses were "more dramatic and more important," she argued. When speaking of girls and sexual delinquency, Johnson spoke not in terms of morality or female purity, but in terms of eugenics and mental hygiene. Unlike a bad boy, she argued, a white girl's "acts are fraught with consequences that are of supreme importance to the race, both socially and biologically, and should have received far more attention from a scientific viewpoint than they have."[9]

The state reformatories under Johnson's charge included two institutions for white boys, one for white girls, and one for African American boys. No state institution existed for African American girls when Johnson arrived as commissioner in 1921, and none was established until 1943, well after her tenure. Recognizing the need for such an institution, the North Carolina Federation of Negro Women, presided over by Charlotte Hawkins Brown, established a private facility for African American girls, the North Carolina Industrial Home for Colored Girls, popularly known as Efland Home, for its location in Efland, North Carolina. Efland Home began accepting inmates in 1925. The mission of Efland Home, according to its 1925 pamphlet, was to "save Negro womanhood and we shall hope to surround these girls with the spirit of Jesus whose memorable words were 'Go in peace and sin no more.'" The school accepted inmates on referral from the North Carolina Board of Public Welfare and the county juvenile courts. The Efland Home board accepted referrals of potential candidates identified as "problems" of the community who exhibited immoral characteristics. Upon acceptance, the county courts would parole the girls to the institution. The school's intended capacity was fifteen inmates, making it quite small compared to Samarcand Manor, which quickly grew to a capacity of more than two hundred white girls.[10]

The nation's first female commissioner of charities and public welfare rested her reform interests on preserving the white race. Yet she showed moderate concern for the states' unreformed black girls as well. Johnson offered tempered enthusiasm and support for reclaiming the state's African American girl delinquents, but found little public support for the cause. Johnson supported the cause and provided Brown with resources, including recruitment of board members, personnel, and staff for the home. Johnson helped Brown secure significant help from the white women in the state, including Fannie Bickett (wife of the former governor, Thomas Bickett) and the North Carolina Federation of Women's Clubs, over which Johnson had once presided. In her 1928 biennial report she recommended that the state take over and enlarge the Efland Home.[11]

Johnson provided help, but in her capacity as commissioner of public welfare it was not much. The state simply had no interest in supporting a reformatory for African American girls. White supremacy reserved the purity of womanhood for white females, and to apply these beliefs to African American girls would counter the basic principle of white supremacy. For white supremacists to build and fund a reformatory for African American delinquent girls would acknowledge the contradictions in perceptions of the purity of white womanhood and threatened notions of white privilege and male privilege. White women who supported a reformatory intended to restore the womanhood of African American girls thus engaged in an activity that detracted from their white privilege. Therefore, the state offered very little support. Despite consistent lobbying by the board of trustees to secure legislative appropriations and recognition as a public facility, the Efland Home received operational funds of only $1,400 to $2,000 from the North Carolina General Assembly each year. In 1939 the home closed, after serving more than two hundred girls, due to inadequate financial support.[12]

Yet Johnson outwardly exhibited genuine compassion and understanding for African Americans. The state, she believed, had a responsibility to African Americans as it did toward whites. Attention to African Americans was crucial, she argued, toward a "better, healthier, and more progressive North Carolina." "The entire state must be interested in seeing that all people in North Carolina live according to at least a minimum standard of health and decency," she argued. She promoted study of social problems in African American communities, proposed aid for the development of community and family programs, and encouraged cooperative self-help programs. In 1925 she established the Division of Work Among Negroes and named a prominent black social reformer, Lieutenant Lawrence A. Oxley, as director. Oxley's reports and activities brought much-needed attention to the social conditions in African American communities. He supported placement of trained African American social workers throughout the state, studied crime through grants by the Laura Spelman Rockefeller Memorial, and frequently visited and inspected public and private institutions that served African Americans. To his supervisor, Kate Burr Johnson, he meticulously reported on operations at state institutions for African American care, including the Morrison Training School (the state's only reformatory for African American boys), the State Hospital for the Negro Insane, the North Carolina Orthopedic Hospital, the Oxford Colored Orphanage, and the North Carolina Tuberculosis Sanatorium. He visited and reported on operations at private institutions, including Efland Home, as well.[13]

When Johnson resigned her position in 1930, African American leaders honored her for her compassion and support. In a surprise ceremony, many "prominent members of the race" representing groups and institutions from across the state, one newspaper reported, presented her with "a beautiful silver bowl with Mrs. Johnson's favorite roses and friezias" [*sic*]. When Dr. Simon Green Atkins, president of Winston-Salem Teachers College, offered a "beautiful tribute to a woman who has given the negroes under her administration the highest hopes," Kate Burr Johnson was moved to tears.[14]

Less interested in restoring the womanhood of African American girls, mental hygienists such as Johnson held out hope to save even the most wayward white girls. In a heart-wrenching account of the story of "Alice," Johnson defended a sexually active white girl who had disgraced her family. Alice, "a quiet, refined, rather dignified" college student of twenty-two had encountered the wrath of her father and brothers when she became pregnant by another college student. When the family discovered her condition, they turned her out for disgracing the family. She bore the child at a maternity home and wanted to keep it. A reformer of an earlier generation might have immediately begun to seek adoption for the child. But Johnson was a mental hygienist who did not condemn Alice for her mistake, nor did she consider Alice a "fallen woman" incapable of restitution. The sexual impulse was normal in girls, she said, although sometimes girls made mistakes. Johnson worried that too many parents, such as Alice's, had so fully restricted and repressed young girls' sexual development and interest in the opposite sex that many girls, in their frustration, turned to delinquency. Generations of abnormal sexual repression, she argued, had caused rebellion among "the youth of to-day." In this social climate, any girl might succumb to a weakness in human judgment. Alice, Johnson noted, had every appearance of a dignified girl. She was not particularly sensuous, hard, or unstable. When Johnson confronted the girl about the reason for her sexual rebellion, Alice cited a lapse in judgment. "Alice, how did it happen?" Johnson queried. "I do not know," was her reply. "I was not myself, it only happened once. I did not love the man. I was engaged to marry someone else whom I still love."[15]

Johnson had argued that the path to prostitution began when families turned unmarried mothers out-of-doors, and she did not intend to let this family commit Alice to such a fate. She hoped to prevent such a tragedy by encouraging Alice's family to support her training and help with the baby's care until Alice secured work. Johnson reprimanded them for considering Alice a "fallen woman" and a disgrace.

> Finally, one night when I had the whole crowd in my house, I faced them. "Now," I said, "I want to ask you a question. Is there a man among you who hasn't had extramarital experience?" They looked at me in silent astonishment. "Your silence condemns you," said I. "Now don't one of you say to me again that Alice has disgraced her family. You are plaguing her into insanity or suicide and making it impossible for me to help her."

After admonishing Alice's male relatives, Johnson issued a defense of women's sexuality. She knew that the long-held assumptions of women's sexlessness stood in the way of Alice's happiness. The "sex impulse," she explained to the crowd of anxious men, appeared equally in both men and women. Despite her ardent attempts to explain her view, she remained incomprehensible to this roomful of men who, she said, clung to the view that a true woman remained chaste and passionless. They held tightly, she said, to the view that "a 'fallen woman' was a 'fallen woman,' a 'bastard' was a 'bastard,' and any family who had either was disgraced and that's all there was to it." In the end, Alice's family triumphed and Alice gave up her baby.[16]

Johnson argued that responsibility for setting white girls back on the straight and narrow ultimately rested with the state. In the story of Ruby, an unmarried twenty-year-old mother, she made the case for state intervention. Ruby's father was an "illiterate preacher," and her mother was a woman of "good mountain stock" who had died when Ruby was young. Rejected by her new stepmother, "the girl was turned loose at an early age just like a calf." According to Johnson, Ruby's "sex delinquencies" began at age fourteen. She dropped out of high school and was a mother by age seventeen. For several years she had lived out west with her brother, who described her as "lazy," "insolent," and "noncooperative." But Johnson saw more in the girl. "The two things we had to build on in Ruby were her love for her child, a beautiful little girl, and her oft expressed desire to take some sort of training." The county social worker made every effort to help Ruby, including placement in a job at the county home caring for inmates. But resources were slim and correctional institutions and children's homes declined her repeated applications for lack of room. Ruby pleaded for help. "I want to go to school worst of all," she pleaded in a letter to a social worker. "Please do all you can for me as I am never going to do any good here. Charles Dean [the reputed father of her child] is always in my face here; temptations are hard to overcome." Help was unavailing and Ruby eventually gave up her baby.

Without state intervention, Ruby descended into a life of immorality. The state had failed Ruby, who, Johnson noted, still "walks the streets." Ruby, she observed, was headed down a treacherous path. She was what one psychiatrist had diagnosed as a "social psychopath, natural prostitute type." Johnson questioned the doctor's classification of "natural prostitute." ("That is debatable," she said). But she accepted the term "psychopath" as "a person under the influence of an abnormal instinct." Intervention might have prevented Ruby's immoral decline. "Those of us who knew Ruby," Johnson noted, "will always wonder whether she would be walking the streets to-day if we could have carried out our plans for her, namely, given her some special training, fitted her into a job, and made it possible for her to keep her baby and be responsible for it." Ruby walked the streets, but Johnson hoped that with further state appropriations other boys and girls could be saved from such a fate. "Prevention is ultimately much more important than remedial care," Johnson observed in a state report. "It is better to discover the boy or girl who tends toward delinquency and change his or her environment so that this tendency may be discouraged, than it is to care for such a boy or girl in the Jackson Training School or in Samarcand Manor after delinquency has developed."[17]

Johnson advocated a southern womanhood that embraced the idea of white women's sexual expression. But there were limits. An unwed mother like Alice who claimed "it only happened once" might be redeemed, but a woman who regularly engaged in sexual liaisons remained suspect. Ruby had transitioned from a good mountain girl to a psychopathic prostitute. Without resources, single mothers faced lives of immorality, interracial sexual liaisons, and the threat of mental defectiveness. State intervention was necessary to save the white stock.

Johnson reasoned that the state could keep women like Ruby off the street by offering financial assistance for raising their children. In 1923 she tapped sympathy in civic, religious, and fraternal organizations, who joined in unison support of a Mothers' Aid bill. It was a new idea—to allow mothers to rear their own children in their own homes, but it had caught on in other states. Most states had begun to pass mothers' pension laws between 1910 and 1920. States generally restricted funds for married women only. Johnson encouraged broader application of the law. In her mind, even unmarried mothers like Ruby should qualify. "With a few exceptions," Johnson noted, "the mothers are capable of 'carrying on' with friendly supervision on the part of the superintendents of public welfare." The legislature made a show of support with a Mothers' Aid law and a $50,000 appropriation. Only two senators dissented. The state promised to match county funds dollar for dollar.[18]

The help was a mere drop in the bucket. Strict eligibility requirements kept most mothers off the list. According to the legislature, eligibility rested upon marital status. Women were to be widowed, divorced, or deserted. Otherwise, the state must find their husband mentally or physically incapacitated or imprisoned. By 1924, sixty-one counties participated, but only 206 families received funds. The state director of Mother's Aid kept quotas low. County welfare workers, though eager to distribute funds, generated waiting lists of approved and pending cases. Nonetheless, Johnson exuberantly praised the program. "It is better to give poor mothers deprived of their husbands' support some such timely assistance as Mothers' Aid than to let such women struggle futilely against poverty with the result that the home must be broken up and the children become institutional charges or dependents upon the community and probably delinquents."[19]

The program likely had little effect on prostitution or adolescent delinquency. It would not have helped the poorest women, such as Ruby. She would not have qualified because her children were born out of wedlock. Few women got help. Of the 206, all were white. Over half were farm families. The mothers' average age was thirty-five, and only one was twenty-two years old, like Ruby. Nearly all had a church affiliation, only 2.5 percent did not. Most women (165) were widowed, but forty-one (20 percent) had living husbands. Most of these had deserted or were imprisoned. The women who benefited generally were economically more stable than Ruby, perhaps because they had once had a husband's support. Many (144 of 206) were employed. Most worked as farmers, cotton mill workers, or seamstresses. A handful had professional occupations, among them a bookkeeper, a clerk, and a "merchant." Where Ruby had no home, sixty of the 206 owned their homes, though many of these were mortgaged. Ruby had one child and no place to live. Mothers' Aid women averaged

3.9 children, 3.9 rooms in their households, and 3.5 beds. Yet the ugly face of poverty loomed. Four mothers had sought help from orphanages, two had sold land, and dozens had begged resources from local community organizations, including the Masons, the Kiwanis, the King's Daughters, and the KKK. In all, 186 cases had received assistance prior to Mothers' Aid. Many mothers sent children to work in the mills, on the farm, and at various odd jobs. And, Johnson noted, many mothers had sold furniture, household equipment, and other important items to stave off poverty and avoid falling to the bottom rung. One mother wrote a letter of deep appreciation. "Without Mothers' Aid checks," she said, "I could not keep my children with me, but would be compelled to give them up. I am trying earnestly to teach them to be Christians, and to love the state that is doing so much for them." These words Johnson took as a mother's implicit commitment to citizenship, words she longed to hear from all the Rubys in the state.[20]

Johnson mourned the inadequacies of Mothers' Aid. Two hundred and six mothers in 1924 was only a small fraction of the women who needed help to keep their children. How else would the state save its troubled girls? Between 1926 and 1928, Mother's Aid helped 542 mothers. The state provided funds only to the legitimate children of "worthy" mothers suffering the loss of a father's earnings through death, disability, or desertion. Johnson explained the difference between a "Mothers' Aid" family and a "Poor Relief" family. The Mothers' Aid family, she explained, "is one that eagerly takes advantage of every opportunity to help itself, while the Poor Relief family holds back from taking any forward steps and is satisfied with a 'hand-to-mouth' existence. The Poor Relief family will never be entirely self-supporting. The Mothers' Aid family will."[21]

Johnson's expansive view of Mothers' Aid reflects her views of womanhood and her white supremacist eugenic agenda. Legislators and some reformers restricted Mothers' Aid to "worthy" mothers—that is, those white and married or widowed whose children were born legitimate. However, Johnson envisioned a Mothers' Aid to all white mothers, regardless of the birth status of the child. To her it made more sense to shore up the whole white race than to reward only the "worthy" few. A state-sponsored campaign to save southern white women from lives of prostitution and eventual degeneracy would maintain the purity of the white race.

For those girls who were not mothers, she encouraged the reformatory model. Situated far from any population center, she reasoned, a reformatory could mold hundreds of unruly girls into the good citizens needed to save civilization. To muster more resources, Johnson decided to hold a meeting. She invited dozens of nationally known reformers and social workers to a conference at Samarcand Manor that would at once cast the reformatory in the national spotlight and bring the best and the brightest minds in the profession to her doorstep. It would be the third annual Conference Held by the Committee on the Care and Training of Delinquent Women and Girls of the National Committee on Prisons and Prison Labor.[22]

She planned the conference for January 26–27, 1922, in Pinehurst, fifteen miles from Samarcand Manor, but unfortunately she picked the week when the Sandhills would experience the worst storm it had seen in years. An uncommonly cold weather pattern settled itself over Pinehurst and neighboring Samarcand Manor. The morning dawned at 19 degrees Fahrenheit. Neither the frigid air nor the overcast sky discouraged participants from arriving the night before, because the snow had not yet begun to fall. Everyone—guests, staff, and especially the girls—anticipated a big snow. The snow began late Thursday night, the first day of the conference. On Friday the snow, three inches of it, covered the earth, grass, and roads.[23]

The snow and the rain that followed significantly altered the program by prohibiting the attendance of the most influential local reformers in nearby Raleigh. The weather prevented the arrival of only a handful of speakers and attendees, but their absence made a difference in the program. Fifty-one guests attended. All of the out-of-towners, who likely had arrived before the storm, were present. Those who did not arrive were North Carolinians who lived less than a day's drive from Samarcand, a trip made impossible on the day of the snow. The original program included a diverse mix of clubwomen, progressive reformers, and mental hygienists, such as Johnson herself. Others invited included county superintendents of public welfare, judges of juvenile courts, and social workers. But the weather had changed the dynamics. Several of the local reformers, including some clubwomen and North Carolina governor Cameron Morrison and Addie Worth Bagley Daniels, wife of the politically influential Josephus Daniels, cancelled on account of the storm.[24]

The absence of the most experienced and politically astute clubwomen and their friends left the program largely in the hands of the mental hygienists, many of whom had national followings. Instead of hearing the evangelical argument of local reformer and Samarcand founder Dr. A. A. McGeachy, the participants instead heard Dr. Katharine Bement Davis, secretary of the Bureau of Social Hygiene. The absence of Governor Cameron Morrison and two other local reformers reduced the only panel about clubwomen's reform efforts from five presenters to the much less-impressive showing of just two speakers. The mental hygienists stole the show, one might say. Captives of the storm heard a New Jersey social work professional Caroline Bayard Stevens Wittpenn, a board member of the New Jersey State Board of Institutions and Agencies, and a Pennsylvanian, the prominent social hygienist Martha P. Falconer.[25]

These unexpected program changes required schedule alterations that changed the direction of the discussion. The clubwomen failed to attend, but the scientists all showed. The vocabulary of medicine and psychology shaped the discussion. Social hygienists encountered little debate because the clubwomen and evangelical reformers simply were not there to counter them. People spoke less of saving "little girls" and "fallen women" and more on reclaiming "defective delinquents" and managing "correctional institutions." The conference opened on a decidedly mental hygienist note. Wittpenn spoke first on the virtues of the "clearing house," a facility where state authorities, she proposed, would perform extensive mental, physical, and social examinations of every convicted girl and woman. According to Wittpenn's model, clearing house authorities would provide recommendations to the court for treatment and sentencing. Upon these recommendations, judges would commit the insane and feebleminded to custodial institutions, the "psychopathic, tubercular, V.D. or other patients" to hospitals for treatment, "and only those capable of being readjusted to society, to reformatory institutions." Another presenter, Dr. Mary B. Harris, reported on "music as a curative." Her report stressed the psychological effect of music and juvenile reform. Girls, she concluded, "respond to the higher type of music and thoroughly enjoy good music that is melodic." She recommended that institutions offer musical programs and study the helpful effects of music on inmates. But, she cautioned, "jazz and music of the lower type has a degrading influence and will tend to demoralize an institution."[26]

Prison science—that is, talk of institutionalization and corrections—also prevailed. Other speakers presented reports on minimum educational standards and special academic courses suited for correctional institutions. More discussion of standardization came from educators who viewed reformatories as training schools for teachers who desired to work with "problem children."

The speakers, chiefly mental hygienists, succeeded in setting the tone of conversation for the two-day, snowbound event. Few people mentioned the "victimization of women" and most conversed in the scientific languages of psychology, diagnostics, treatment, and standardization of care. They spoke in the language of mental hygiene. The old language of reform was the vocabulary of the past.[27]

Social workers used the word "feebleminded," the most common and most ill-defined term to describe a condition or sickness that indicated a mental deficiency. Kate Burr Johnson certainly used the word with great frequency. She employed it as if it were a clinical term to describe psychological conditions that she and other mental hygienists defined as "mentally sick and deficient." During the early twentieth century, social workers and scientists had begun to use the terms "feebleminded," "mentally defective," and "psychopathy" as medical explanations for women's criminal behavior. Johnson, an advocate of prison reform, freely used these words to explain that correctional institutions were ineffectual because they focused only on retribution and not on inmate reform or mental health. "Mental deficiency or sickness," she warned, "is the most serious of all social problems, and . . . mental hygiene is one of the most important social measures." Corrections authorities should recognize that "our criminal and delinquent population is largely made up of the mentally sick and deficient," she argued. In fact, she and other mental hygienists sounded the alarm: the feebleminded plagued society both inside and outside prison walls. One authority, the director of mental health and hygiene, reported that fifty thousand North Carolinians had inherited subpar intelligence.[28]

Mental defectives, Johnson argued, fell into three categories, all requiring state care at varying degrees. In identifying those persons in the first two categories, the "fairly well adjusted" and the "partially adjusted," state professionals might offer strategies of "adjustment" to compensate or to cure the mental sickness. Most mental defectives were "fairly well adjusted" to society, Johnson explained, and presented no great cause for concern. Many others were only "partially adjusted," and required help. She argued that the state should identify "partially adjusted" kids at an early age and employ trained professionals in reformatories and the public schools to train or "adjust" them for society. Johnson asserted that "our public schools are filled with mentally deficient children whom we are trying to squeeze, like square pegs into round holes, into programs made for normal children." State authorities, Johnson argued, must find these children before they presented a social problem. The traditional school environment, she stated, would flummox these children, who would only grow resentful of conditions too difficult for their "inferior abilities" to meet. They required special education classes composed of "ungraded class rooms," and with trained teachers in charge "many of these children can be successfully trained and prevented from becoming delinquent and institutional cases."[29]

In other words, Johnson held out hope for "partially adjusted" kids. Public schools could save many of them. Reformatories offered hope for the most troubled of this class. Reformatories were not prisons, she observed, but correctional schools designed "to make good citizens out of the boys and girls." The state had funded Samarcand and three other reformatories, two for white boys and one for African American boys, to address delinquency. No one ever intended these reformatories to care for the most feebleminded and mentally defective. The reformatories handled "maladjusted" individuals, trained them, and readjusted them for society. No inmate would stay indefinitely. The reformatories were designed for those "neither feeble-minded nor psychotic," Johnson observed, "but who need careful and detailed study to determine the particular habit or character defects which cause their maladjustments."[30]

Those worst off and the greatest threat to the race, Johnson argued, constituted "a third group of mentally unfit," who were "entirely unadjusted" and required segregation from the rest of society. Johnson estimated that two thousand North Carolinians were "idiots and imbeciles who are burdensome and unable to care for themselves, and the delinquents." Institutional training and parole might socially adjust half of this group. The rest, those most feebleminded of the mentally unfit, required permanent institutionalization. Segregation would protect the larger population from contact with the most deficient. Johnson adamantly opposed any social contact between the mentally unfit and all other children, whether normal or "far below normal," in society. "The presence of feebleminded children with normal children is a handicap to the normal children," she asserted, "in that such feeble-minded children not only require a large amount of attention that should go to furthering the development of the normal children, but also in that these feeble-minded children fail to offer the normal competitive stimuli that are so necessary to bring out the best in the normal children." To protect normal children, the state had equipped one institution, Caswell Training School, for these most mentally defective.[31]

So Samarcand Manor was a house of refuge for the most hopeful cases. At least that was the intent. The trouble with Samarcand was that many of the inmates appeared to exhibit profound feebleminded characteristics. Johnson unhappily noted that intelligence tests completed at the institution indicated that 55 percent of the girls at Samarcand possessed "subnormal intelligence" that suggested a "relation to the delinquency of these girls." She allowed that not all the girls were "feebleminded," but "in all probability cases of dementia praecox, while others are perhaps only psychoneurotic." Apparently Johnson had accepted these other two classifications, "dementia praecox" and "psycho-neurosis," as defects less serious than "feeblemindedness." Johnson did not want to suggest that a majority of girls at Samarcand were among the "mentally unfit." To do so would suggest that Samarcand served not as a reformatory for wayward girls, but as an institution for the mentally insane. Yet, she concluded, "the low intelligence quotient is of significance as an indication of an abnormal mental condition."[32]

How did Johnson test for feeblemindedness in the early 1920s? After World War I, social work authorities embraced the idea of intelligence testing of inmates in public facilities. North Carolina had administered intelligence tests to many of the girls at Samarcand between 1922 and 1924. In all, authorities tested 487 people. Two hundred and thirty-four were girls at Samarcand Manor, and the 253 "other cases" presumably served as a control group, though the study offers no other information about these cases. Subjects took the Terman General Intelligence Test, an I.Q. test named for its inventor, psychologist Lewis Terman, who adopted the view that intelligence was hereditary. Terman had coined the term "I.Q." for "intelligence quotient," a ratio of a child's mental age to his chronological age times one hundred. He had adapted his test in 1916 from the Binet-Simon model used to assess normal as well as mentally deficient children. In 1916 intelligence testing sparked controversy. One judge had refused to admit a Binet-Simon test as evidence and remarked that "standardizing the mind is as futile as standardizing electricity." But after World War I, when the army widely adopted the use of I.Q. tests to measure the intelligence of draftees, the practice took off in the civilian population. I.Q. tests typically consisted of two versions. Examiners usually administered "alpha" tests, a series of word analogies, arithmetic problems, synonym-antonym puzzles, and common sense queries. A beta test offered a pictorial version for non-English-speaking subjects. Critics argued that testing outcomes favored subjects with strong scholastic skills and reflected the race, sex, and cultural background of the individual tested. The Samarcand study noted that as a general rule an "intelligence quotient of less than 70 or 75 indicates feeblemindedness." The study concluded that fifty-four girls were "feebleminded" or "probably feebleminded" and another thirty-four girls were classified among eleven variants of mental disorders, including "senile dementia," "psychoneuroses," and "alcoholic deterioration." The study classified only four individuals as "normal." How did examiners define "normal"? The study offers no answers. In fact, examiners enigmatically followed the term with a question mark. [33]

In intelligence studies, puzzling results were par for the course. But Johnson believed it was better to be safe than sorry. One might never know where insanity might lurk. I.Q. tests offered a window into the psyche, and Johnson believed only psychiatric and mental hygiene experts possessed the diagnostic tools and understanding to rule on the mental capacity of every individual in society. It is absurd, Johnson contended, to have a jury, a group of laymen, rule on an individual's insanity and then determine his or her sentence. It was far too dangerous to the public welfare to rest this responsibility with laymen. A board of scientists, such as a Mental Hygiene Board, should make recommendations on insanity pleas after careful study of the prisoner in question. Any real headway in treatment and correction, she concluded, must include mental health and hygiene as basic to reform. "Mental deficiency or sickness is the most serious of all social problems," Johnson asserted, and "mental hygiene is one of the most important social measures." The problem represented such a severe threat to society, Johnson concluded, that the state should assume the responsibility for testing the mental capacity of every individual. In an effort to see this promise through, Johnson employed her power as commissioner of charities and public welfare to create a Bureau of Mental Health and Hygiene and to name Harry W. Crane as director. Crane was responsible for the study and control of mental defectives in the state institutions, and he reported directly to Johnson. Together they addressed the question of measuring feeblemindedness. Created in the first year of Johnson's tenure as commissioner, the bureau conducted its first mental exam on October 14, 1921. [34]

Passing judgment on mental defectives was no easy task. Even Johnson had trouble with it. One day she visited a pretty young girl who sat in jail, arrested for "immorality." The commissioner had hoped to gain some insight into her mental condition. She was fifteen or maybe sixteen years old and had none of the marks of the streets on her. "She sat and talked to me as quietly, intelligently and pleasingly as any of your daughters might have done," she told a group of physicians. The girl intrigued her because Johnson had heard rumors of her terrible behavior, a wicked temper such that "language vile beyond imagination pours from her lips." She was violent, too. The welfare officer "will carry a scar to his grave," Johnson noted, where she tried to kill him in her unprovoked rage. For this attack she served her jail sentence, and her time had nearly expired. What are we going to do with her? Johnson inquired rhetorically to her audience. To turn her loose on society was a great injustice, but no expert could say with certainty that she was insane. "Neither can I," Johnson confessed. Yet the law should allow that a psychiatrist observe her, "for in her we have a certain prostitute and a potential murderess on our hands." With proper treatment, she might be saved or restrained, thus to save the public from her antisocial acts. [35]

Left untreated, she argued, mental sickness evolved into a deficiency that passed to offspring and weakened the gene pool. In her speeches and reports, Johnson often cited a case study of the Fehlers as exhibiting all the classic behaviors of inherited deficiencies. The study had extended back several generations. Johnson explained that all proper names were fictitious. State authorities had adopted the name "Fehler," the German word for defect, to indicate that family members were "defective socially, morally, mentally, and in a great number of cases physically." In the early nineteenth century, Ed Fehler, a "fairly good citizen," settled in Sand County with his wife, Hannah, who had an "unsavory reputation" and was rumored to have an extremely violent temper. The couple populated Sand County with 260 direct descendants, many of whom lived their "nonproductive lives in misery, poverty and ignorance" or "crime and degeneracy." Intelligence testing provided evidence of the "very low average mental development" of thirteen family members. Legend described the men as mean, violent, and crazed. A son, Red Fehler, was notorious. At middle age he went insane. Neighbors knew him to rove the woods, climb the trees, and howl like an animal. Rumor attributed nineteen legitimate and "countless illegitimate" children to another son, Tom Fehler. "The story goes," Johnson recounted, "that Tom kept so many women who bore his children that in order to keep account he cut notches on a walnut stick for each one." The women were of a low order, too. Nan Fehler, Tom's daughter, bore four illegitimate children, "one of which," Johnson reported, "was a mulatto." Apparently the Fehler women reveled in sexual promiscuity. Eleven Fehler women bore a total of forty-four illegitimate children. "Born in poverty and misery," Johnson stated, these were "white children living with their illegitimate mulatto brothers and sisters."[36]

Johnson's story of the Fehlers appeared in the state-published biennial reports. The story served as her evidence why North Carolina had to actively seek to protect its gene pool by controlling women's sexuality. Social workers interpreted the Fehler case study as an example of sexual immorality run amok. Sexually promiscuous men created great problems, but the greatest danger occurred when white women engaged in illicit interracial sexual relations. Their nonwhite offspring not only inherited their mothers' genes, but threatened to pass their undesirable traits through more of the white race. Mental sickness, made biologically endemic over the course of generations, threatened the strength of the white race by the spread of poverty, ignorance, crime, and degeneracy. Men went insane or immoral. Women engaged in unchecked promiscuity that threatened endless interracial intermixing as "white children" lived with their "illegitimate mulatto brothers and sisters." Women engaged in interracial sexual relations had passed the point of no return. They were maladjusted, had lost the traits of southern womanhood, and could not be restored. Unless North Carolina found a solution for isolating these women, Johnson feared, the white race would never be the same.

Johnson argued that large families of mental defectives presented a grave problem to society and required a radical solution. Segregation offered an option but, according to Johnson, was ineffective. Permanent institutionalization of such large family networks wasted taxpayer dollars and did not offer appropriate reproductive control. Another viable option, Johnson asserted, was sterilization of the unfit. Johnson viewed sterilization as a legal and humane method of protecting society from mental defectives. The U.S. Supreme Court had upheld a Virginia sterilization law in the case of Carrie Buck, whose guardian had sought to prevent her sterilization in a public institution. Justice Oliver Wendell Holmes ruled in favor of the procedure. "Three generations of imbeciles are enough," he declared. Johnson agreed that in some cases sterilization was appropriate. "I do not recommend sterilization as a cure-all," she cautioned, "nor of much value as a general remedy, but as a great help both to the individual and the public in specific cases."[37]

North Carolina had passed its first sterilization law in 1919, but the law displeased Kate Burr Johnson. It yielded unsuccessful results because there was simply too much red tape for it to do any good, she complained. The 1919 law applied only to institutionalized patients. When medical staff at an institution recommended sterilization, the chief medical officer first had to seek approval by a sterilization board, the governor, and the State Board of Health. Johnson herself had for years lobbied legislators to lift these requirements to allow more authority to an institution's medical staff. In her first biennial report as commissioner she recommended statutory revisions that would allow surgical procedures on feebleminded inmates in institutions when approved by two physicians, including one representative from the State Board of Health, or the secretary of that board. She got her wish when the General Assembly passed new, more far-reaching legislation in 1929. The new law allowed an institution's "governing body or responsible head" to make the decision based on the best interest of the patient or "for the public good." County commissioners also could request sterilizations for mentally defective or feebleminded residents not in a public facility. An individual's guardian or next of kin could petition for their beneficiary's procedure. The law required the state to seek a properly trained surgeon and to file a detailed medical and family history on the patient. Finally, to Johnson's pleasure, the law named a new review board that included herself as commissioner, the secretary of the State Board of Health, and two chief medical officers of a state insane asylum. Petitions no longer required the inconvenience of awaiting the governor's approval. To provide for greater convenience, the law replaced his signature with hers.[38]

So we return to the central question raised at the outset of this chapter. How did Kate Burr Johnson reconcile her strong beliefs in eugenic policies such as sterilization with her experience and faith in social reform? Framed yet another way, did she emphasize nature or nurture as the cause of juvenile delinquency? In other words, did hereditary deficiencies, that is, nature, cause delinquency? Or did dismal social conditions, that is, nurture, produce unhealthful conditions that caused children to "go bad"? Historians have shown that eugenics advocates drew from both biological determinism and environmentalism. Some eugenicists argued that mental inferiority was biologically determined. These eugenicists wanted to segregate mental defectives from society and sterilize them to prevent their reproducing more of the "unfit." Other eugenicists adopted a more Lamarckian approach. Poor environment that caused delinquency eventually produced mental degeneracy, which then passed on through hereditary decay. By the environmentalist argument, mental defectives might reintegrate into society after sterilization. Either way, eugenicists believed that once mental degeneracy was introduced to the line it could not be reversed.[39]

Johnson adopted a Lamarckian philosophy to better conform to her ideals of southern white womanhood. Mental degeneracy threatened the purity of the white race because genetic mutation, she argued, developed from constant exposure to a bad social environment. Among whites who suffered negative social conditions, early intervention might correct delinquency. However, once entrenched, the behavior produced degeneracy and no amount of social reform would prevent its inheritability. Johnson advocated controlling women's sexuality and reproduction as the solution. She argued that the state could protect the remaining normal genetic pool by segregating mental defectives and, at the very least, sterilizing them before reintroduction into mainstream society. Once sterilized, mentally defective women who lacked the pure strains of southern white womanhood no longer posed a threat to the race.

Johnson, an earnest advocate of the poor, was a woman of deep compassion, sympathy, and understanding for the most unfortunate residents of the state. But history must also remember her as being motivated by her firm racial predilections. As an officer of public welfare she served the state's poorest, but she also was a mental hygienist who feared the genetic inferiorities of the defective and feebleminded, both white and black. At once she sought to shape the state welfare board into one that would control the sexuality of white girls and improve conditions in African American communities. While her sympathy for African Americans is clear throughout her reports, her racial justifications lie there as well. Whites could not ignore the realistic truth that though theoretically separated from African Americans by Jim Crow custom and legislation, they nonetheless lived alongside them. "The close proximity of the Negro community to the white community in most cities and towns of North Carolina," wrote Johnson in her 1928 biennial report, "makes the social, health and sanitary conditions, with their hazards, questions of vital concern to the entire state."[40]

In answer to the question, "Did nature or nurture produce juvenile delinquency?" Kate Burr Johnson accepted both without contradiction because of her unusual view of southern white womanhood. Poor environment led astray even the most genetically pure white girls, who, without proper reclamation, would engage in sexual behavior that would lead to a tainted racial stock. Her role as the chief state public welfare officer, she argued, was to preserve the natural superiority and purity of the white race through the nurturing of white girls, the icons of white purity, by reform and reclamation. In the worst cases where the racial stock was already tainted, she accepted sterilization as an extreme measure to protect the genetic purity of the race.

5

A Modern Girl's Dilemma

GIRL RUNAWAYS AND SEXUAL ANXIETIES
IN JAZZ-AGE NORTH CAROLINA

AN UNREPENTANT MARGARET ABERNETHY watched as pieces of hair fell like ragged shards into her lap. The shears, blunted by use, no longer produced an even cut. It was the second time Mrs. Bradshaw had cut her hair for trying to escape. Miss MacNaughton had ordered the haircut to humiliate her. Her hair, once worn in a simple bob like the other girls', was cut to the skin. From her appearance, she hardly seemed capable of much trouble. Margaret was sixteen and her shape was just beginning to flower. She was small, thin, and blonde. According to one source she had a "bad complexion" due, perhaps, to the diseases of childhood, such as measles and smallpox, known to produce scaring on their victims. In demeanor she was "timid," "frail," and "very pathetic." Her sadness and low self-esteem most certainly stemmed from an unhappy childhood. A shaved head certainly would not improve her self-esteem problem. Margaret knew that she would become the butt of all the jokes by the girls in Chamberlain Hall. Boys were not allowed on the grounds of Samarcand Manor, except on visiting day, when brothers could visit their sisters. Perhaps because of this exclusion, the inmates took great pleasure in ribbing any girl who looked as though she did not belong.[1]

This chapter examines how Margaret Abernethy and other white girls of the jazz age negotiated adult expectations of their behavior. The jazz age had introduced into southern society several variants on white womanhood based upon iconic national stereotypes of women. In an effort to define their behavior and their place in society, white girls of every class precariously straddled the iconic symbols of the 1920s: the provocative and sexualized image of the flapper, the growing independence of the "new woman," and the traditional idea of the pure and chaste southern lady. This chapter will explore these symbols, with particular attention to their southern meanings, and examine adolescent white girls' identification, use, and manipulation of these symbols, particularly in reference to runaway behavior. All adolescent white girls, those of every class, struggled with these competing and very contradictory images of womanhood. While middle-class girls usually expressed their sexuality and independence within acceptable cultural parameters, girls from the working class, like Margaret, endured a higher degree of scrutiny because their public behavior on the streets posed a greater threat to the public racial and social order.[2]

Margaret Abernethy had arrived at Samarcand with a clean record. Except for her predilection toward flight, she had not committed any crime. How, then, had she come to be there? The Lenoir County authorities sent her to the reformatory after convicting her father, Dewey Abernethy, a Kinston plumber, for the crime of incest upon Margaret. Margaret received no sympathy whatsoever from her stepmother. The girl's parents had separated when she was a toddler. When she was four, Dewey Abernethy took her from her mother. Shortly thereafter, her mother died, her father remarried, and her stepmother raised her, though reluctantly. When Margaret turned ten, her father began to visit her bedroom at night. The incest occurred two or three times a week— whenever her stepmother left the house. The crime continued for three years. Margaret reported that he drank, but usually he was not drunk when he sexually assaulted her. She told no one. He had threatened her into silence. "It was against my will." Margaret reported, "he kept me scared." One night her stepmother caught him in the act. She "had just had a baby and he slipped down to my room," Margaret remembers, and she caught him in Margaret's bed. Margaret vividly remembers what happened next:

> After she found out, my daddy told me he was going to kill me, and I ran away and went to a lady's house not far off. They got the police and carried him to jail. Then they took me to the county home. I didn't like it there and I would run away.

Local authorities sent her to a children's home in Smithfield. But upon her father's conviction, social workers sent her to Samarcand.[3]

As Margaret's case attests, one did not have to commit a crime to land in Samarcand. Many of the girls, even those considered the very worst, began their criminal history as runaways. Some were girls who rebelled against the rigid conventions of womanhood. Others, like Margaret, turned to the streets to escape a bad situation at home. Why were girls incarcerated and punished for the mere act of running away? Though "running away" did not constitute an official crime, it nonetheless posed a serious offense. Girls who "ran away" or "ran around" unsupervised represented a threat to the white supremacist notion that idealized all white girls and women as southern ladies. White elites maintained power through the idea of "white purity," a concept that rested on the perceived chaste behavior of white girls and women. When adolescent girls behaved in "modern" ways at work and in leisure, they alarmed southern white elites, who saw their sexualized behavior as a threat to "white purity," the cornerstone of whiteness.[4]

Margaret must have posed that threat, because for her crime the worst was yet to come. The runaway could expect a whipping, a punishment reserved for the most incorrigible girls. This beating would be her fourth in the twenty-nine months she had lived at Samarcand. She knew the drill well. Miss Stott, or perhaps Mrs. Bradshaw, would perform the whipping with a hickory switch. Always Miss MacNaughton was present. The superintendent would select six girls, usually other incorrigibles whom she desired to teach by example, to hold her. Margaret, dressed only in a thin cotton gown, would lie face down on the bare floor of Chamberlain Hall with her hands above her head to receive her due. Generally the whipping consisted of six to ten strokes and lasted two or three minutes. But, according to Miss MacNaughton, that number depended upon the girls' "attitude or reaction toward punishment," and the staff kept no records of the beatings.[5]

Because it was her second escape attempt, Margaret probably endured a more extreme beating. Viola Sistare, a registered nurse, saw some of the physical injuries on girls who sought treatment at the Samarcand hospital. The first girl came during Sistare's initial month on the job. She came immediately after the whipping, Sistare reported, "with her buttocks and upper limbs *very badly* bruised, black and blue with large welts." Sistare had no authority over discipline, nor did she wish to intervene. But the child could not sit, and she found it difficult to walk. So Sistare followed her professional calling and treated the girl. Then she released the girl to her cottage. The nurse later learned from another teacher that the girl had stopped at another cottage, where some of the teachers asked to see the marks. "The child showed them," Sistare learned, "and one teacher fainted." Occasionally the authorities called upon nurse Sistare to provide "a pill" to settle a girl's nerves. One night, Sistare herself delivered the drug to Chamberlain Hall. She arrived just in time to witness a whipping.[6]

They would not have whipped a southern lady. Admired, treasured, revered, the southern lady represented one of the more iconic and long-lasting images of the South. She was the symbol of purity and grace that white society expected its girls to imitate and its boys to admire; its women to emulate and its men to covet. Before the Civil War the most desirable southern lady was modest, chaste, kind, gracious, and appealing to men. After the war she retained these attributes but was also resourceful, smart, and assertive, especially in defense of white southern ideals and social conventions. The southern lady served as a political symbol, too. White supremacists invoked the southern lady in their campaign to deprive African Americans of their right to vote. The end of slavery, they argued, allowed African American men uncontrolled access to white women. White supremacists constructed a rallying cry for the protection of "white womanhood" from the black male rapist as the chief justification for depriving African American men the right to vote.[7]

At the turn of the century, middle-class white women invoked the southern lady as public housekeeper to justify their move into government and politics. African American women used the image, too. White supremacists had denied any respect for black womanhood. But African American women appropriated the image of the southern lady for themselves to justify their own public participation and club organization. As southern ladies, they worked to improve standards for African Americans in their communities. They called this process "racial uplift," and the African American middle class considered it their moral obligation. Middle-class white women adopted the language of "sisterhood" and, in theory, embraced poor white women and working-class women as ladies. Put in practice, they often viewed women of lesser advantage as dependent on their protection and enlightenment. Thus, while social convention required that Margaret Abernethy act like a southern lady, her class standing did not require Samarcand authorities to treat her like one.[8]

Margaret came of age in an auspicious era, a time when the southern lady met the "new woman." The new woman, college-educated and frequently single, had iconic status by the time Margaret arrived at Samarcand. At Margaret's birth in 1915, the new woman, in the guise of personalities such as Jane Addams, Florence Kelly, and Kate Burr Johnson, had fundamentally shaped the progressive reform and settlement house movement. In Margaret's toddlerhood the new woman vigorously campaigned for suffrage, and by the time Margaret turned six she had won women the right to vote. The new woman made strides in the South as well. In the person of Kate Burr Johnson, superintendent of the North Carolina Department of Charities and Public Welfare, she assumed significant public roles not only in reform but also in government. As an adolescent, Margaret would have recognized the strides of women such as Nell Battle Lewis, a prominent newspaper columnist and female attorney, who exuded poise and confidence in her words and deeds. Sometimes, the new woman and the southern lady were one and the same. Leading suffragists in the South often emerged from the ranks of the privileged class and were raised with a sense of *noblesse oblige*. As reformers, suffragists expressed their humanitarianism and showed faith in government action. Their contacts with suffragists in northern and western states also led them to view feminism favorably. Not all southern suffragists favored a federal amendment. Some, such as Laura Clay and Kate Gordon of Kentucky, advocated state-by-state reforms that would exclude African American women from the right to vote. Furthermore, for every woman suffragist there existed a woman antisuffragist. Thus it is true that the new woman rejected the traditional conventions, but in the South she nonetheless retained many of the white supremacist assumptions that were at the heart of the southern lady ideal.[9]

The rise of the new woman in the 1920s also led to the emergence of another iconic symbol of independence, the "flapper." While the new woman symbolized financial independence, the flapper exuded sexual independence. Indeed, the flapper represented both an image and an attitude. A visual ideal of the flapper rests in American literature, namely F. Scott Fitzgerald's "The Opposite Pirate." In this popular short story the flapper protagonist, Ardita Farnham, was rich, young, and spoiled. "She was about nineteen, slender and supple, with a spoiled alluring mouth and quick gray eyes full of a radiant curiosity." Fitzgerald described her as more an ornament than a person. In clothing she was "adorned rather than clad." Nonchalant, disdainful, and scornful, the flapper, Ardita, tempted with her beauty, flirted insatiably, and displayed a fickle and capricious nature, particularly with men. Fitzgerald defined the "flapper" derisively. "A man could not trust a flapper," he wrote. Even Ardita disliked the nomenclature. When the dashing Tobey Morehand twice called her one, she finally responded, "I wish you wouldn't call me that."[10]

Ardita, the flapper, also represented a spirited independence akin to the new woman of the 1920s. Where the new woman was educated, financially independent, and self-confident, the flapper took the new woman to the extreme. Yes, she was independent, but more in spirit than in finances. Impetuous, brash, defiant, sexually confident, she at once appealed to men and drove them crazy. When her family insisted that she marry, F. Scott Fitzgerald's Ardita rebelled by joining her uncle on his yacht. Fitzgerald understood the gendered double standard a girl must transgress to become a flapper. He demonstrates this double standard in an exchange between Ardita and Morehand about the virtues of courage. Ardita explains that courage comes in many forms. In men, courage brought strength. It inspired the "Bloody prize-fighter" to come up for more. For her, courage inspired defiance to disobey her family's wishes. Courage meant rebellion and adventure. It is "the liking what you like always; the utter disregard for other people's opinion—just to live as I like always and to die my own way." Her courage was really a defiance that contributed to her sexual appeal to men. As she spoke to Carlyle, she flaunted her physique in a series of perfect dives off the edge of a cliff on a small island south of Florida. But courage also meant that Ardita could express herself sexually and with confidence of self that women of past generations could not: "My courage is faith—faith in the eternal resilience of me—that joy'll come back, and hope and spontaneity. And I feel that till it does I've got to keep my lips shut and my chin high, and my eyes wide—not necessarily any silly smiling. Oh, I've been through hell without a whine quite often—and the female hell is deadlier than the male."[11] F. Scott Fitzgerald understood that in the 1920s, courage meant something different to women than to men.

Had Margaret Abernethy heard about flappers? Lack of written information means we cannot know for certain. Most likely she had. Fitzgerald wrote about flappers in 1920 when many middle-class American women considered showing leg and bobbing hair. The term "flapper" first applied to a teenage girl with a gawky frame. By the end of the 1920s the term came to designate young women in their teens and twenties who embraced daring conduct in speech, dress, and behavior. By the time of Margaret's internment in December 1928, the flapper had gone mainstream. Margaret, having completed fourth grade, might have read the short stories or novels of F. Scott Fitzgerald, who popularized the notorious type in literature. She might have discovered the short-haired, leg-baring, dance hall-visiting, make-up-wearing image in *Flappers and*

Philosophers (1920), a collection of short stories that included the famously scandalous "Bernice Bobs Her Hair."[12]

A more likely avenue by which the flapper reached most girls was through movies. The flapper image was perhaps one of the greatest marketing devices of the Hollywood film industry. Actress Colleen Moore titillated moviegoers in *The Flaming Youth* (1923), described by one reviewer as "Intriguingly risque but not necessarily offensively so. . . . The flapperism of today, with its jazz, neckerdances, its petting parties, and its utter disregard of the convention, is daringly handled in this film. And it contains a bathing scene in silhouette that must have made the censors blink." By the end of the decade, Moore, Clara Bow, Louise Brooks, and Joan Crawford typified the glamorous flapper in movies such as *Painted People* (1924), *The Perfect Flapper* (1924), *Flirting with Love* (1924), *Naughty but Nice* (1927), and *Our Dancing Daughters* (1928).[13]

The glamour industry quickly capitalized on the flapper's commercial appeal. Some studies from the period showed that young women who went to the movies paid close attention to the stars' appearance and behavior. Corporations, such as Lucky, Chesterfield, and Kenilworth Cigarettes, Pond's Cold Crème, and Coco Chanel fashions brandished the flapper image to appeal to the spending power of young working women. In Atlantic City, businessmen dared controversy by producing the first Miss America beauty contests to lure tourists to the beach in the winter.[14]

Expansive glamour ads monopolized the newsprint in North Carolina. Lux Toilet Soap advertised in glamour ads that appeared in Raleigh's *News and Observer* in 1929. "Nine out of 10 screen stars use Lux Toilet Soap" it alleged in a quarter-page advertisement featuring photographs of sixteen famous young female actresses, including Clara Bow and Joan Crawford. The advertisement pronounced that "Blondes-Brunettes-Red-heads: . . . *all* screen stars alike have the initial appeal of smooth lovely skin." Ad executives had sorted the photos by each actress's hair color, noted her studio affiliation, and included a caption that featured a quote by each young star. The ad listed another eighty-nine young female actors by hair color as "a few more of the host of lovely stars who guard their skin this way." Marion Daves, whose photograph appeared most prominently in the top left corner, was "famous for her blonde loveliness, says: 'smooth skin is a great asset. I am delighted with Lux Toilet Soap.'" Samarcand girls—that is, those who lived in the honor cottages—probably knew of many of the actresses on the list. The staff allowed girls to earn credits for good behavior, rewarded monthly by free admission to film showings at local movie houses. Film and advertisements, not to mention the controversies surrounding the Miss America pageants of the 1920s, likely appealed to the girls in the reformatory as much as to girls who lived at home.[15]

This panoply of images—the new woman, the southern lady, and the flapper—must have left adolescent girls dazed and confused about societal expectations. The southern lady acted demure and chaste, the new woman appeared intelligent and engaged, and the flapper behaved rebelliously and sensually. The contrasts were sharp. However, a keen-eyed observer will discover at least two of the images creatively reconciled in the pages of the local newspaper. There, middle-class women reinterpreted the southern lady by incorporating features of the flapper. They did so in a manner acceptable to even the most scrutinous critic. Raleigh's *News and Observer* frequently published photographs of "Brides and Brides-to-be," lucky ladies who struck poses akin to their more glamorous counterparts in the film industry. Resplendent in fashionable clothing, hats, jewelry, furs, and an occasional feather boa, young North Carolina graced the Sunday morning society page. Styled bobs, carefully applied rouge and lipstick, and come-hither smiles beckoned readers with flapperance charm. Yet these women were not girls of the street. Engaged or recently married, the women of the society page were already spoken for in the traditional southern lady sense. Captions that accompanied the pictures affiliated them with their husbands and rarely identified their given maiden names. One would not know that Mrs. Garland Gilbert Wooland was also known as Miss Judy Van Dyke except by reading the accompanying article that dutifully listed all the brides, bridegrooms, and their parents. Thus, brides and brides-to-be at once could strike a flapper pose and remain within their fathers' and new husbands' protective association.[16]

The city of Raleigh and local corporate sponsors repurposed this middle-class flapper pose for a corporate event promoting Raleigh and its local industry. In February 1929, the *News and Observer* and the North Carolina Theater sponsored a production called *Our American Girl*. The film would depict "Raleigh, its people, locale, and many interesting and beautiful sports." It would feature Ruth Perkins, who, the *News and Observer* indicated, "will soon flash out of the sky and come to a halt at one of the local landing fields." The corporate sponsors already had arranged to include two young Raleigh women in the production, Dorothy Bogue and Anne Moore. Upon Perkins's auspicious arrival, the three girls would enjoy a tour of Raleigh, but not alone. Escorted by an "entourage" that included the staff of the film company, the girls "will take in all of the chief industrial plants, civic centers and points of natural beauty for miles around," the *News and Observer* announced. The paper promised a most exciting spectacle. Promotional photographs depicted Moore and Bogue in glamorous poses and an article pledged that they would "be costumed in the latest fashion" by a local department store named Boylan and Pearce. Described as "young ladies" or "pretty Raleigh girls," Moore and Bogue at all times retained their propriety. Provided that she had a proper guardian to escort her from one event to the next, a young woman could safely experience the excitement of the modern era.[17]

To mimic a flapper and remain a southern lady was a middle-class privilege. In a forum as public as the newspaper, women could strike a flapper pose, provided they did so as a daughter, wife, or wife-to-be. They could enjoy a glamorous lifestyle, but only under the auspices of a guardian or corporate "entourage." Wealth was not necessarily a prerequisite for glamour. However, a woman required resources, enough to allow her to resume her role as a southern lady once the excitement waned. Yet for a brief moment she experienced a degree of freedom that the flapper symbolized.

These three icons, the southern lady, the new woman, and the flapper, left girls few alternatives. Women did not have the choice to dress or behave as men. If they did so, they would be arrested. At least, that was the message given by the authorities and by the *News and Observer* when the paper reported the arrest of nineteen-year-old "Billy Cain." Cain, despite any misunderstanding produced by her given name, was a girl. She had moved to Durham from Akron, Ohio. On Friday January 18, 1929, patrolmen arrested her on a charge of "nuisance" after she had "successfully concealed her femininity under the habiliments of man for many weeks," according to the newspaper. Cain was found guilty and sentenced to jail, where, the journalist concluded, she "will have the opportunity of wearing the clothes of her sex, for the next 90 days, at least." When questioned why she dressed like a man, Cain kept mum. She refused any statements to the judge, would not question the officers who had taken the stand to testify against her, and adamantly refused any comment to inquiring reporters on the courthouse steps. Apparently Cain had been in the company of three young men, who also were arrested, also on the charge of nuisance. One was released and two others were sentenced to ninety days' work on the county roads. Authorities had arrested all four young people for equal unruliness, and on the same charge. But the *News and Observer* relished the story of Cain's arrest in a tawdry headline, "Arrest Girl Found Wearing Men's Garb: Sentenced to 90 Days in Jail on Charge of Being a Nuisance." The headline ignored the boys' transgressions and boldly indicted Cain for violating sexual conventions, apparently the greater crime.[18]

As Billy Cain's story suggests, girls who found refuge or adventure on the streets created significantly more anxiety among authorities, parents, and the press than boys who did the same. Sometimes girls chose the streets over abuse or violence at home. When girls believed that conditions at home or at school had become too restrictive, they simply ran off. Mary Emily Gardner, a fourteen-year-old girl from a tiny hamlet called Bear Grass, located in Martin County, left not to seek glamour but to escape bad conditions at home. When she reached Greensboro she found refuge with an elderly couple and felt safe enough to write a few letters home. She ran, she said, because her mother whipped her. She had no intention of returning home and, she warned, "any attempt to bring her back would prove useless." She planned, she said, to find employment in a "five-and-ten-cent store in Greensboro."[19]

Social workers at North Carolina State College linked running away to neurotic behavior and instability among white girls. In the 1920s, some social workers had begun to turn away from social explanations of behavior to psychoanalytic diagnoses. In questions of child guidance and delinquency, these social workers advocated an individualized reform plan to "readjust" the child rather than a social reform of the child's environment. Lena B. Ladu and K. C. Garrison adopted this theory in their 1931 study of the "mental ability" and "some temperamental traits" of white delinquent women and girl inmates in North Carolina's penal institutions. In December 1931 they published a study of 138 inmates, including Margaret Abernethy and the other fifteen arson defendants at Samarcand Manor, Central Prison, and the North Carolina Farm Colony for Women. They first applied a Dearborn Group Intelligence Test. The IQ test indicated that as a group the girls were "retarded one year mentally." They followed the IQ test with a questionnaire known as Thurstone and Thurstone's Neurotic Inventory, which included questions related to feelings, happiness, lonesomeness, feelings of worthlessness, "Ideas of Persecution" by family members, desires to run away, and suicidal tendencies. A high number of positive responses indicated an "emotionally unstable personality." A low number of positive responses indicated "an absence of emotional strains or worries, and thus indicates an emotionally well adjusted personality."[20]

Ladu and Garrison argued that delinquent girls, particularly those charged with incorrigibility, running away, and sex offenses, had the highest degrees of emotional instability. They concluded that white girl delinquents in their study were "social misfits." Girl delinquents differed from girls in the general population by having more intense conflicts with family members, liking to be alone, "[tiring] of people quickly," and having difficulty making friends. They had abnormal fears of lightning and fire. "The delinquent girls in the penal institutions in North Carolina," they concluded, differ from the control group of girls "in their stronger desires (a) to be alone a great deal, and (b) to run away from home; (c) in their marked feelings of temper, (d) misery, and (e) persecution; and (f) their abnormal impulse to commit suicide." They concluded that the girls should be classified as *"emotionally maladjusted"* or "fell in the range of subjects who *should have psychiatric advice.*" Ladu and Garrison provided little information on their control population, a group of college freshmen who tested "emotionally well adjusted" or "unusually well adjusted."[21]

Thus Ladu and Garrison concluded that delinquent girls, including runaway girls, possessed some sort of personality dysfunction. Her inability to adjust to society provoked family conflict and suicidal tendencies. Society, the press, and academics classified girls who ran away as emotionally unstable, maladjusted, and requiring psychiatric advice. Runaways did not possess any of the traits of the obedient southern lady, the confident new woman, or the buoyant flapper. They were social misfits who required psychiatry and, in cases of intense family conflict, institutionalization. Unlike the control group of mostly middle-class college girls presumed to be well-behaved, confident new women, the runaway by virtue of her own personal dysfunctions could not meet any of the expectations of white womanhood.

Girl runaways created great consternation and angst. When a girl disappeared, parents, authorities, and journalists created elaborate theories of abduction or seduction to save her from the reputation of a delinquent. When Mary Emily Gardner disappeared, everyone suspected a local man named Will Knox of kidnapping her. Authorities arrested Knox but dropped the charges when Gardner's letters surfaced, showing that she had left on her own accord. In Asheboro, parents and authorities began to suspect abduction when two fourteen-year-old girls went missing one week following the disappearance of two older girls, Moline Bulla and Thelma Luck, all from the same community. Witnesses reported that Virginia Hammer and Kathleen Amick had run off with Enoch Nelson, described as an eighteen-year-old boy driving a stolen Buick. The newspaper suggested an insidious motive by further describing Nelson as "tall and very dark," a language that evoked white supremacist fears of African American men preying on young, white women. Yet three days later, authorities discovered the girls in Washington, D.C. Enoch Nelson, for his part, had returned to Asheboro to "give himself up." The newspaper this time described him as a "young white man," as if to vindicate him for their earlier suggestion that he was of another race. Authorities charged him with car theft and placed him under a $500 bond, but they levied no charges of kidnapping the girls.[22]

Paradoxically, reports of foul play proved more acceptable to parents than explanations of teenage mischief. Parents and authorities conjured up scenarios of abduction and kidnapping to prove their daughters' innocence and save them from a bad-girl reputation that might despoil their ladyhood. It seemed as though North Carolina was experiencing a rash of disappearances in the winter of 1929. Between January 21 and February 21, the *News and Observer* reported the disappearance of ten girls between the ages of ten and nineteen. Eighteen-year-old Alice Haynes and fifteen-year-old Florence "Dorothy" Younger, described as "high school students of good families," caused great anxiety in their parents, who proceeded with a multistate search. When, after two days, authorities had turned up nothing, their parents reluctantly began to consider a kidnapping theory. They did not entertain the idea for very long. Shortly thereafter relatives found the girls at the home of Jim Malone, a Sarasota newspaper employee who had once lived in Greensboro. Apparently the girls had decided to stage an unexpected visit to Malone and his family when they learned that they had failed to advance to the next grade in school. The girls, carrying only a small semiautomatic pistol, two cans of soup, a box of crackers, and overnight bags "covered with college stickers" apparently had hitched rides all the way to Florida. The North Carolina relatives quickly dispatched Florence's father to retrieve them.[23]

While parents considered kidnapping theories, the *News and Observer* delighted in conveying the girls' mischievousness. Often journalists spun stories of girl runaways in lurid flapperesque terms. A flapper certainly might have secured only a pistol, soup, and crackers in a hasty attempt to find adventure.

Most certainly she would have dressed well. When fourteen-year-old Virginia Hammer and Kathleen Amick disappeared, the witnesses who last saw them described them as if they were dressed to the nines. The girls wore "tan coats trimmed with tan fur," reported the *News and Observer*. Readers also learned that Virginia Hammer had "long bobbed hair, dark brown and very curly." Kathleen Amick wore her "very black" hair in a "short bobbed style." The bobbed-hair-and-fur-coat observation not only evoked Fitzgerald's literary descriptions but also resembled the images that appeared on the society pages. On January 20, 1929, Moline Bulla and Thelma Luck, described in a newspaper account as "good looking 17-year-old school girls," disappeared from Asheboro. Local reports had placed them on a south-bound bus en route to Hollywood. Later, Moline Bulla and Thelma Luck turned up in Jacksonville, Florida, employed as waitresses in a hotel.[24]

Boy runaways garnered nowhere near the attention. Reports of missing boys or young men made the paper only in the most dire circumstances. Such was the case when the three Higgins brothers (Jesse, Elliot, and Randolph) decided to take Fred Englert, a friend from New York, duck hunting in the Brunswick County swamps. They left in a boat on the morning of Friday, January 18. When they did not return, their distraught father, T. L. Higgins, coordinated an intensive air and land search that continued for weeks. A brutal environment for hunting ducks or dead men bubbled up from the Brunswick County swamps. Semiprecious cypress shaded both terrestrial and aquatic surfaces for searches from the air. The ground campaign had it no better. The centuries had built up layers of wet, thick sediment that could entrap a man looking for bodies. Bears, wolves, and occasional alligators posed inconveniences as well. On the second day of the search, a rescue plane piloted by a German war ace crash landed. The pilot and two passengers (including a deputy sheriff) survived but with multiple fractures and head and body lacerations.[25]

Tragically, all four boys had drowned. In detailing the progress of the search, the *News and Observer* floated the idea of foul play when a gun and empty shells belonging to one of the boys turned up near a whiskey still hidden in the swamp. The news media later reported that all four had drowned. A Coast Guard cutter, the *Medoe,* had discovered the young men's capsized boat. The media's theory was affirmed not long thereafter with the discovery of a canteen, two lunch tins, and Randolph's coat near the boat. T. L. Huggins still had not found his three sons when searchers recovered the first body, Fred Englert, three weeks after the mission began. Two days later, African American fishermen pulled Randolph Huggins, age sixteen, from the Cape Fear River. The father identified the remains of his youngest son from the deck of a yacht that had accompanied the fishing skiffs. The devastated Higgins could not bear to look upon his deceased son's face, but recognized him by his clothing. The body of his eldest son, Jesse, age twenty-six, had floated five miles downriver before its discovery two weeks later. Finally, on February 23, an African American fisherman retrieved Elliott's corpse opposite the Caroline Beach Pier, fifteen miles downriver from Wilmington. The coroner finally confirmed accidental drowning, despite the sensationalist treatment of the incident in the press.[26]

Girls understood the risk to their reputation if caught "running around," even if no sexual intercourse occurred. Some girls who voluntarily ran away actively manipulated a kidnapping charge to deceive authorities and preserve what they could of their reputations. Fourteen-year-old Lottie Inge experienced second thoughts after a joyride in April 1931. One evening she had set out on a walk to visit her aunt when a car carrying two men pulled up alongside her. The men invited her for a ride, and she accepted. The joyride lasted until dawn. As morning approached, Lottie began to grow anxious, unsure how to explain her behavior to her family. She decided to fabricate a kidnapping story and enlisted her companions' help. The three drove to the railroad yard and located an abandoned boxcar. There, at Lottie's request, the boys bound her with ropes and left her in the car. Then the boys disappeared and Lottie Inge began to cry for help. At some point the empty boxcar began to move. A railroad worker had hooked the car to a train bound for Selma. Inge, now genuinely afraid of what would become of her, redoubled her efforts to make herself heard. The force of the coupling knocked the door open, and a brakeman, seeing something "rolling around in the car," gave the signal to stop the train. By the time the men discovered her, she was weak and exhausted from screaming. The men delivered the girl to Rex Hospital, where doctors examined her and determined that she had not been sexually assaulted. She told the police that the men had forced her into the automobile at gunpoint and then tied her up in a boxcar, where they had left her all night.[27]

Neither the police nor the media bought the whole story. In a desperate attempt to save her reputation, Inge continued the fiction for days. She peppered the story with details, such as the manner in which they hoisted her into the car and bound her, accounts of their many "improper advances" toward her, and a description of their intended sexual assault, foiled only by a train whistle that frightened them away. However, the details kept changing and the authorities suspected fraud. Her mother, who professed that Lottie never lied and "paid little attention to boys," stood by her daughter's account. Lottie's friends at school wrote a letter testifying to her "sweet and loving disposition," her honesty and truthfulness. They also called upon the town fathers, as if in substitution for her own who had died when she was young, to "make a nation-wide search for the guilty parties and bring them to justice and save our girls from such terrible things." Two days later Inge confessed that she had faked the kidnapping. She had persuaded the two men to tie her up and leave her in the boxcar, where the brakeman found her not long thereafter. She said she was afraid to return home after the joyride and concocted the kidnapping as a ploy to save her reputation. Perhaps because Lottie's father was dead and she did not live with her mother but boarded elsewhere, her reputation was particularly vulnerable to ruin from one night on the town.[28]

If the frequency of these reports is any indication, people must have thrilled in reading about girl runaways. In 1928 and 1929, the nation must have faced an epidemic, or it feared so. Even Hollywood tapped the angst. By late 1920, the film industry had labeled flapper-like behavior as sexually delinquent. A 1928 film, *Port of the Missing Girls*, eroticized the image of female adolescence and presumed that only the most vigilant parent possessed the power to re-strain her from sexual delinquency. One ad for the film asked, "75,000 girls in the past year were reported missing. Why do girls leave home? Where do they go? Who is to blame?" Another source estimated the girl runaway figure at eighty thousand, and another suggested that girl runaways posed "one of the biggest problems" facing law enforcement.[29]

The anxiety over girl runaways was part of a significant shift occurring in attitudes about adolescence. In the "jazz age," progressive reformers and psy-chologists typified female adolescence as a period of sexual restlessness and irrepressible thrill-seeking that sometimes produced wayward results. Adoles-cence, argued G. Stanley Hall, represented a series of stages of childhood where the mind had not fully matured along with physiological changes in the body. The process was necessary for mature adulthood, but was marked by conflicts, emotional distress, and rebellion. While rebellion was considered normal, the behavior of "bad" adolescence evoked particular concern. First noted among working-class youth, rebellious behavior had spread to middle-class teenagers by 1920 to form a new "youth culture" distinct from adult values and practices. Youth exhibited a new boldness with public intimacy, sexuality, dating, and supervised recreation. Parents feared the "girl problem," once attributed to poor working-class girls, in their own daughters' sexual behavior. The state of North Carolina viewed Margaret Abernethy as one of these problems. She had first run away when her father threatened to kill her. Her stepmother had had her committed to Samarcand after her subsequent flights from welfare and county homes. [30]

In Chamberlain Hall, Samarcand Manor's discipline dorm, Margaret was in the company of girls of the same habit. Parents and state courts had committed many girls for the same offense, "running away," or like charges such as "running around" or "incorrigibility." Of the sixteen girls charged with arson in 1931, seven were incarcerated for committing no other crime but "running away." Another three had arrived at Samarcand on charges of "running around" unsupervised alone or with other girls and boys. Some girls had runaway histories so flashy their stories could have made the headlines of the day. Mary Lee Bronson ran away to Richmond with another girl and a man she did not know. Gone a week, she found refuge at the Richmond YWCA. Her father, a traveling salesman, and her mother, a bookkeeper at a bakery, threw up their hands and had her committed to Samarcand. At age fourteen, Edna Clark ran away three times. Described as quite attractive, of medium height and spirited in personality, she possessed an uncommon charm, polite manners, and keen interest. Apparently six months of working in a Halifax cotton mill at $7.50 a week had sparked her wanderlust. Her mother recovered her after the first episode, an unannounced trip to Kinston to visit a girlfriend. But problems continued at home. Her father drank and quarreled often with her mother, who possessed no high reputation. Soon thereafter, Edna turned up in Raleigh, where she had been living with a man for a month. This time her father retrieved her. After her third escape, police caught her with a girlfriend and two boys in possession of liquor. Her father gave up and had her committed for "not going to school." Josephine French, another cotton mill girl, landed in Samarcand for "running around" and being "immoral and disobedient," according to her mother, who had her committed. [31]

"Running away" was one common denominator. Girls who landed at Samarcand also shared another common trait. Half of the girls charged with arson by the state in 1931 arrived at Samarcand at their parents' or a guardian's behest. Their parents would not or could not restrain their behavior. Sixteen-year-old Chloe Stillwell was committed by her parents after she ran twenty miles from Kinston to New Bern and stayed alone in a hotel. Her father, a carpenter, and her mother requested her asylum at Samarcand since she was "in danger of prostitution." Another seven girls had experienced the tragedy of the death of one or both parents. Margaret Abernethy's experience ranked as one of the worst; her mother had died in Margaret's toddlerhood and her father served prison time for sexual assault upon Margaret. Four more girls had witnessed the death of both parents. Five girls knew one living parent and six knew both. In some cases the parents were estranged. Rosa Mull's mother resided at the Western State Hospital for the Insane at Morganton. Pearl Stiles's father made a living as a farmer in Canton during part of the year and as a bootlegger in Washington, North Carolina, during the rest. Parents worked in low-wage occupations that, in 1931, would not have escaped the darkest hours of the Great Depression. They were farmers, cotton mill workers, butchers, and carpenters. Only Mary Lee Bronson had parents who worked in jobs—sales and bookkeeping—remotely considered professional. Apparently her parents had lost the patience and resources needed to raise a new woman or a proper southern lady.[32]

Thus, very few of the girls at Samarcand came from families of the middle class, and those who did came from the middling sort. They fit none of the conventional stereotypes—southern lady, modern woman, or flapper. Their identities likely rested on none of these. However, their perceptions of self may well have centered in another realm of women's experience—that of the cotton mill girl. During the agricultural crises of the 1920s and 1930s, adolescent girls from rural North Carolina flooded the cities to find jobs in the textile industry. In some of the mills, women accounted for 30 percent or more of the labor force, of which 75–80 percent were single females age sixteen to twenty-one. These young working women thus transformed the work and recreation environments of the textile industry. The cotton mill communities encouraged social recreation among women, who created their own homosocial relationships that drew upon popular forms of amusements and commercialization of the larger youth culture. Their distinctive working-class friendships, sexuality, and protest styles helped lead to a moral panic among southern white elites.[33]

State records provide no firm data on the number of girls at Samarcand who came from cotton mill communities, but references suggest a significant minority of girls came from this background. Records show that at least two of the sixteen suspected arsonists had worked in a cotton mill and another two had mothers that were textile employees. In a 1931 investigation, Superintendent Agnes MacNaughton commented on illiteracy rates among the general population of inmates at Samarcand, which she linked to the mills. Over 25 percent of the inmates could neither read nor write, she reported, "this high degree of illiteracy is found to be among the girls who have been working in the mills." Samarcand was a frequent destination for girls from juvenile courts in cities in the Carolina Piedmont, so it is very possible that textile workers and their daughters, perhaps as much as 25 percent of the inmates, helped shape the culture at the institution. [34]

The textile industry, despite its historically repressive working conditions, provided women and girls with a certain degree of economic and personal freedoms that women's traditional role of domesticity did not. Female cotton mill workers grew accustomed to earning a wage and living in boarding houses, or finding a home in a mill village while their husbands worked on the farm. While employees paid women workers historically meager wages, women nonetheless gained a buying power in the marketplace and contributed to their family's financial resources. Annie Viola Fries remembers her family's financial status in her childhood as relatively good. While she didn't have much in the way of material goods, she contrasted her situation with the very poor, who she described as "really right down dirt poor" and "lived in little old shacks." [35]

But the mills were also a place of protest and militancy, especially during the 1930s. Employers cut pay, refused workers any bargaining rights, and instituted the stretch-out, a technique that speeded up the production line and eliminated workers' meal and bathroom breaks. As conditions worsened in the 1920s and 1930s, female employees engaged with their male peers in protest by participating in strikes and union activity. Jacquelyn Dowd Hall argues that women employees, at the center of this protest, expressed a gendered militancy that reveals working-class women's and girls' identity. "Through dress, language, and gesture," she argues, "female strikers expressed a complex cultural identity and turned it to their own rebellious purposes." During the protests, women strikers taunted guardsmen and mocked men's egos. They wore men's clothing, used aggressive language, and adopted a swagger in public protests and demonstrations. The press and authorities linked their labor militancy to sexual misbehavior, which sometimes led to charges of promiscuity or prostitution. [36]

With ties to the mills, histories as runaways, and records of sexual misbehavior, the Samarcand inmates represented the quintessential "disorderly
women." Even the staff noted their proclivities towards "sex perversion." When
staff members complained that girls practiced sodomy and masturbation, the
superintendent responded that the reports were "probable." Superintendent
MacNaughton explained that in most correctional institutions sexual activity
is common and difficult to control. Among girls who learn early of sex relations, "as have the majority of inmates of Samarcand," MacNaughton argued,
"there is a strong probability that the more experienced and strongly sexed she
is the more inclined she will be to seek sexual gratification in whatever manner
she can even while an inmate of the institution." Perceptions of promiscuity
persisted even though many of the girls were victims of rape and incest. Like
the textile strikers whose labor militancy was mistaken for promiscuity, the
Samarcand inmates' high incidence of sexual activity and venereal disease led
to charges of "sex perversion." The reputation of sexual misbehavior among
Samarcand inmates thus resonated with concerns about adolescence, labor
militancy, and the larger "youth culture" of the jazz era. Middle-class girls
could engage in modern amusements by reconciling them with southern womanhood, but the working-class or reformatory girl could not. Her very disorderliness in public threatened the idea of "white purity," the heart of southern
womanhood and white supremacy.[37]

Margaret Abernethy's own disorderliness had led her to a head shaving,
and now she anticipated another whipping. Having failed at escape, Margaret
sought another way out. One night in March 1931, she and another Chamberlain Hall girl named Marion Mercer acted on a possible opportunity. Somehow,
at approximately 6 P.M., Bickett Hall had caught fire. When the news spread,
Margaret and Marion were in the Chamberlain Hall dining room with the
other girls. The Chamberlain Hall teachers feared that the fire might spread.
As a precautionary measure, the teachers ordered the girls to empty Chamberlain Hall of all movable items. Perhaps it was their view from Chamberlain of
Bickett aflame that inspired Marion Mercer and Margaret Abernethy to initiate
the next act in a sequence of events. "We got all the things out of Chamberlain,"
Abernethy later confessed, "and Marion Mercer and I were taking things out
of the teacher's room and we decided to set Chamberlain on fire." Another girl,
Ollie Harding, overheard the conversation and offered to get matches. "But
she didn't get them," Abernethy bragged, "I got them." Then she and Marian

Mercer climbed into the attic of Chamberlain Hall. They brought along a stocking, the only thing they had that might light quickly, and "Marian lit it and we put it in a hole," Abernethy remembered. The stocking burned and the girls ran down from the attic. But this was not the fire that would destroy Chamberlain Hall. Teachers noticed this fire at about 8 P.M. and extinguished it before any damage occurred. "I thought that if I set it on fire they would send me out, and I was tired of that place." Margaret later confessed. The young girl was completely unaware that she had committed a capital crime. Not more than an hour later, Chamberlain Hall caught fire again, and this time it burned to the ground.[38]

FIGURE 15. Hope Chamberlain Hall. In 1931, fire destroyed Chamberlain and Bickett Halls at Samarcand Manor. Sixteen girls were arrested and tried on charges of arson, a death penalty offense. Samarcand Manor School Records, Mars 5899, Department of Public Safety, State Archives of North Carolina, Raleigh, North Carolina.

FIGURE 16. The parlor in the Hope Chamberlain Hall. Each hall had a parlor where, some reports state, staff forced girls to whip others, or else be whipped themselves. Samarcand Manor School Records, Mars 5899, Department of Public Safety, State Archives of North Carolina, Raleigh, North Carolina.

FIGURE 17. Bickett Hall, burned in 1931. Samarcand Manor School Records, Mars 5899, Department of Public Safety, State Archives of North Carolina, Raleigh, North Carolina.

FIGURE 18. Grace M. Robson, the mental hygienist who replaced Agnes MacNaughton, established the classification and parole system that evaluated girls for sterilization. She served from 1933 to 1943. Records of Samarcand Manor, Division of Adult Correction and Juvenile Justice, Department of Public Safety, Samarcand Manor School, Eagle Springs, North Carolina.

FIGURE 19. Girls did all the work at the institution, including laundry, cooking, and work on the farm. These girls, dressed comfortably in overalls, are pitching hay circa 1922. Records of Samarcand Manor, Division of Adult Correction and Juvenile Justice, Department of Public Safety, Samarcand Manor School, Eagle Springs, North Carolina.

FIGURE 20. In 1929, Vera Bush, Sophie Melvin, and Amy Schechter, three union orga-
nizers at the Loray Mills in Gastonia, were among sixteen arrested and charged with the
murder of the police chief during the Loray Mills strike. Girls at Samarcand with ties to
the mills likely heard of the women. "I came to Gastonia shortly before the fatal raid to
organize the children to help in the conduct of the strike," Sophie Melvin told the *Labor
Defender* in 1929. Image from Fred Beal, "Who Are the Gastonia Prisoners?" in the *Labor
Defender* (September 1929), Nell Battle Lewis Collection, PC 255, box 30, North Carolina
State Archives.

FIGURE 21. Editorial cartoon of the death of Ella Mae Wiggins. Labor union journals circulated among cotton textile workers. Girls with ties to the mills may have seen or read these publications in their communities. This political cartoon from the *Labor Age* depicts the 1929 death of union organizer Ella Mae Wiggins. Image from "North Carolina," *Labor Age* (October 1929), Nell Battle Lewis Collection, PC 255, box 30, North Carolina State Archives.

FIGURE 22. Electric chair, Central Prison, 1929. From 1910 to 1930, North Carolina executed no women. The state reserved its death penalty for African American men. The state utilized the electric chair at the Central Prison in Raleigh from 1910 to 1938. The chair occasionally appears on exhibit at the North Carolina Museum of History. Samarcand Manor School Records, Mars 5899, Department of Public Safety, State Archives of North Carolina, Raleigh, North Carolina.

FIGURE 23. Some of the Samarcand inmates, including the arson defendants, were cotton textile workers or children of cotton textile workers. Their identities tied to the mills, they likely held sympathy for the strikers and patterned their behavior after "cotton mill girls." Here, cotton textile workers and their family are evicted from a North Carolina home, circa 1929. Nell Battle Lewis Collection, PC 255, box 30, North Carolina State Archives.

ARSON DEFENDANTS AND THEIR COUNSEL

Some of the fourteen Samarcand girls on their way to the Moore county court at Carthage yesterday when twelve of the fourteen were sentenced to the State Prison for indeterminate sentences of eighteen months to five years for burning two buildings at Samarcand are shown above. One of the defendants is quite obviously taking the last final puff of her cigarette before facing the judge for sentence.

At the left is Nell Battle Lewis, Raleigh attorney, who while representing only one of the defendants, assumed the brunt of the defense in a dramatic presentation to the judge. It was Miss Lewis' first major case since she started the practice of law and in it she undertook to show unsatisfactory conditions in the State correctional institution.

Twelve Samarcand Girls Get State Prison Terms

FIGURE 24. In the wake of the trial, the Raleigh (N.C.) *News and Observer* printed this article, which includes the only known image of the arson defendants, at right. Their attorney, Nell Battle Lewis, is pictured at left. At the conclusion of the trial, the judge sentenced twelve girls to five to eight years imprisonment at the Central Prison, suspended the sentences of two girls, and discharged two girls. The Nell Battle Lewis Collection, PC 255, box 30, courtesy of the North Carolina State Archives.

FIGURE 25. "Our Three Youngest." This promotional photo appeared in the Samarcand Manor *Biennial Report, 1928–1930*. Samarcand Manor housed girls under ten until Superintendent Grace Robson stopped the practice in 1934. Image from Samarcand Manor *Biennial Report, 1928–1930*, Nell Battle Lewis Collection, PC 255, box 29, North Carolina State Archives.

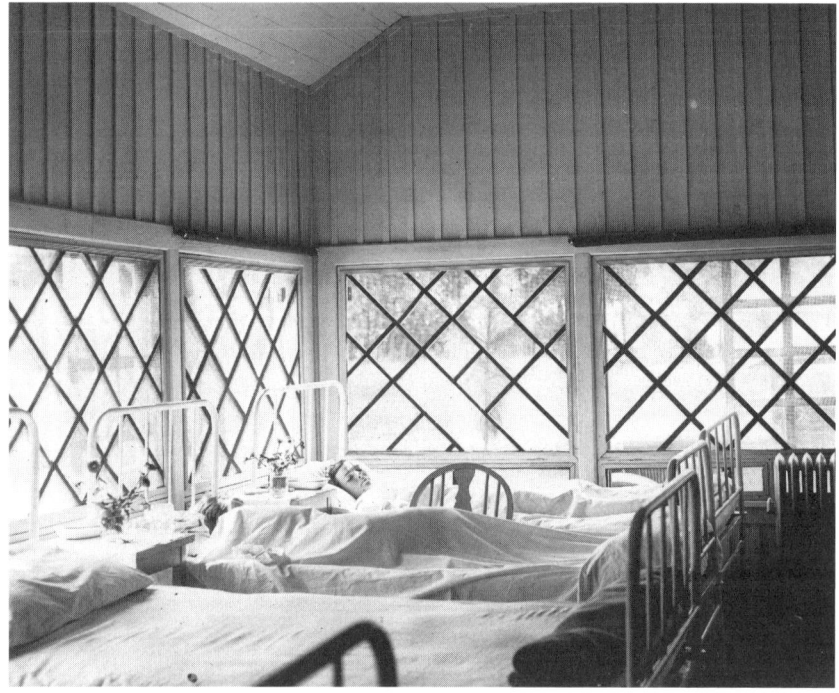

FIGURE 26. Hospital infirmary. The Samarcand Manor infirmary always had served as a center for quarantine, isolation, and treatment for venereal disease. In 1934, the infirmary became the first link in the chain of decisions that determined a girl's sterilization. Samarcand Manor School Records, Mars 5899, Department of Public Safety, State Archives of North Carolina, Raleigh, North Carolina.

6

Not Penitent Yet

THE STRATEGY OF THE DEFENSE IN
THE SAMARCAND ARSON CASE

NELL BATTLE LEWIS, NEWLY RETAINED as counsel for the defense, had just learned that her clients reportedly had provoked a riot on Thursday, April 30, 1931. And that afternoon, six of the sixteen alleged arsonists had set fire to a mattress and created a disturbance in the Moore County jail while awaiting trial on capital punishment charges for the March 12 burnings of Chamberlain and Bickett dormitories at Samarcand Manor in Eagle Springs, North Carolina. Newspapers reported the incident as "riotous" and "rebellious." They nicknamed the girls "fire bugs" and, more notoriously, "fiends incarnate." The *Moore County News* called it a "mutiny." Reportedly, the girls had "set fire to their cell bunks, smashed every window pane they could reach, which was practically all of them, and finally attacked members of the local fire department who came to their rescue, with pocket knives." The *News and Observer* described the incident as a "rampage" that ended only when the girls were subdued by a fire hose turned upon them. It was their second attempt to burn down a jail. The same girls previously had caused even greater damage at the Robeson County jail in Lumberton. Authorities moved the perpetrators to Moore County while the other, apparently more docile girl defendants remained in the Lumberton jail.[1]

This chapter examines the strategy of Nell Battle Lewis in her defense of the suspected Samarcand arsonists. Lewis, a well-known example of a southern new woman, had embarked upon her first case. She was already an expert and published author on the social problems posed by capital punishment in the South. As a well-known columnist for the *News and Observer*, she most certainly understood the degree of public exposure the firebug articles would generate. As an attorney she would have known that the media reports would produce negative perceptions of the defendants in the impending jury trial. Her strategy was quixotic. Early on, she devised a consistent and focused defense of the girls as wayward adolescents, budding into womanhood but neglected by the state. Throughout the trial, all but one girl professed innocence, and Lewis characterized them all as victims of negligence in a state institution. However, in the last minutes of the trial and just as the judge began to consider sentencing, Lewis suddenly changed course and pursued a modified version of an insanity defense: the girls set the fires, she said, because they were feebleminded adolescents and victims of their environment. Her last-minute shift to an insanity defense at once reflected her inexperience in the courtroom and contradicted her public relations campaign to characterize the girls as rational, normal teenagers. It is not clear if Lewis's end-of-trial shift reflected a sudden loss of faith in the girls or, perhaps, a last-ditch attempt as a neophyte lawyer to lessen their sentences. However, it did mark a dramatic departure in the representation of the girls as normal, white adolescents. By introducing evidence of psychological tests and expert testimony of the girls being feebleminded, Lewis reinforced the idea that these girls had not met proper expectations of womanhood and classed them as mental defectives. To Lewis, the shift may have seemed well-intentioned and innocuous, but the public consequences were significant, because in 1931 the state was rife with social unrest and hostility, especially toward female workers in the textile mill communities where some of the girls had roots.

Lewis had joined the legal defense team just days before the girls' jail rampage. George W. McNeill, a prominent local attorney, headed the defense team. McNeill, a seasoned veteran of the courtroom, had practiced law for a quarter century. However, he had never before appeared in court with a female attorney. He could not have chosen to partner with a more high-profile woman. Though it was her first case, Lewis had passed the bar several years previous. She had achieved the status of cause célèbre for her advocacy of numerous progressive reforms, including women's rights, women's suffrage, due process for the mentally unfit, and prison reform. She had written on the textile strikes and union activity in North Carolina, and she had pursued numerous research projects in social welfare. In 1929 she collaborated with Public Welfare Commissioner Kate Burr Johnson and Lawrence A. Oxley, director of the Division of Work Among Negroes, to produce a treatise titled *Capital Punishment in*

North Carolina that provided the history and statistics of the death penalty in the state. She was a neophyte in the courtroom, perhaps, but she seemed the perfect fit for the Samarcand arson case. Her writing focused on the most serious and controversial social and political issues of the day. Yet the public knew her best for her weekly contributions to her *News and Observer* newspaper column, "Incidentally."[2]

Initially, Lewis was hired to represent just one of the girls, sixteen-year-old Virginia Hayes of Rocky Mount. In the first few days after she took the case, Lewis began an investigation that seemed like it might produce an angle favorable to her client. The attorney-client interview revealed that Hayes was no notorious offender. She had been committed to Samarcand after running away from home. Hayes told Lewis that she hated Samarcand and would rather remain in jail than return to the reformatory. Lewis learned from her client that at Samarcand they "just treat you terrible," "gave us horse beatings," "make the girls fire the furnace," and, probably due to overcrowding and a severe infestation of bed bugs in the discipline hall, the girls "had to sleep on one blanket on the floor." Furthermore, a medical condition promised to elicit her some sympathy from the jury. She allegedly had participated in the first fire at the Lumberton jail, but a severe case of appendicitis placed her in the hospital and away from the Moore County jail during the more highly sensationalized jail rampage of April 30.

Hayes denied any responsibility for the first Samarcand fire. She claimed that during the first outbreak she was "out in the yard" with the other girls. During the second conflagration at the school, the one that eventually destroyed Chamberlain Hall, she said she was in bed, apparently among one of the five to fifteen girls (accounts differ as to the exact number) that the attendant had locked in their rooms for running away. It appeared she had a tight alibi, provided Lewis could verify her claims. Locked in her room and unable to cause any stir, she seemed the quintessential victim wrongly accused. "Miss Clay," she told Lewis, "would not open [the] doors of rooms in which 15 girls were locked, but gave other girls the keys to unlock them." It was a wonder the poor girl had escaped unharmed.[3]

Hayes's case seemed open and shut except for one problem: authorities claimed that she had confessed to the crime. Within hours after the first conflagration that took Bickett Hall, Superintendent Agnes MacNaughton ordered all Chamberlain Hall residents into the reception room of Tufts Hall, a nearby honor cottage that had escaped damage. Nearly four dozen girls with no place else to go crowded into the neighboring dormitory. "The first thing I said to them," MacNaughton later professed to Lewis, "You must remember that if you had nothing to do with the fire to stay out of this." Any girls having "anything to do with [the fire], she ordered, must step into the teachers' reception room." At once, fifteen of the forty to fifty Chamberlain girls that stood before her went into the next room. The sheriff arrested these fifteen girls and took them into custody.[4]

On further inquiry, Lewis discovered that the authorities, including the sheriff of Moore County, had failed to follow basic principles of due process in arresting the girls for arson. Of the fifteen girls, none were questioned individually or privately. MacNaughton admitted that her follow-up to the confessions was sparse. Several of the girls, she remembered, had said, "I did it," and MacNaughton's chief concern was to secure the grounds from further danger. "All I was thinking about was to get the dangerous ones off the grounds," she told Lewis. Authorities the next day arrested a sixteenth girl, Pearl Stiles, the only defendant from Bickett Hall, based not upon her own confession but upon the testimony of another girl in her dorm.[5]

The night's confusion also clouded memories, so that no consensus existed on the number or identities of teachers who witnessed the purported confessions. MacNaughton named three teachers and staff as present: her secretary Estelle Stott, Edith Marksham, and Edna Ross, matron of Chamberlain Hall. Ross, she said, had spoken to the girls. However, Edna Ross later told Lewis that she was not present at the questioning. It was Stott, not Ross, who claimed that twelve girls had confessed to "setting fire" to the buildings and three had confessed to "knowing" about the fires. Estelle Stott verified that she was present at the confessions and claimed to have warned the girls "that it [arson] was a capital offense and that they should be careful." However, MacNaughton must have missed this vital exchange because she claimed to be unaware that arson was a capital crime in North Carolina. How could the superintendent have missed Stott's exchange with the girls? Was she absent from the room? Were events so chaotic that Stott's words never reached MacNaughton's ears? If so, what other exchanges might have been missed or gone misunderstood? As if to confuse matters further, Maude Moore, the matron of Bickett Hall, who no one placed at the scene, claimed that the girls were individually questioned. How could Moore have firsthand knowledge if she were not at the scene?[6]

Thus, the "star" witnesses, Agnes MacNaughton, Estelle Stott, Lillian Crenshaw, Edna Ross, Maude Moore, and Lillie Bradshaw, had presented the prosecution with two troubling contradictions. First, no one could agree on how the girls had confessed, whether as a group or individually. Second, no one seemed to know precisely which teachers and staff were present nor what was said to the girls to elicit their confessions.

Yet the girls' lives rested on the sorting of the details. The law recognized a big difference between the act of arson and the mere playing with matches. The states' legal definition of arson rested upon the English Common Law, specifically meant to secure the habitation or "house" from "the malicious and voluntary burning" by another. Arson, like burglary in the first degree, was considered "an offense against the security of the habitation." If committed maliciously upon a dwelling house or sleeping apartment of another, it necessitated the threat of the death penalty. Long-established legal principles held this law in place, despite a brief interlude when North Carolina's Reconstruction legislature lifted the death penalty in 1869, only to have it restored by the Redeemers two years later. Arson, because of its threat to human life, differed from "burning" or "setting fire to" any other property, such as public buildings, schoolhouses, or churches. North Carolina specified that "the offense that carries the death penalty is one against the habitation only and one that will greatly endanger human life." By contrast, the willful and malicious burning of a jail or other county building was not defined as arson but did constitute a felony necessitating five to ten years of imprisonment. In any case, an indictment of arson required not only establishing intent, but also evidence of burning defined as the "charring of wood." The prosecution thus had strong justification for arson in the Samarcand case. Two dwelling houses, Chamberlain and Bickett Halls, not only were charred, but burned beyond use. All the prosecution needed was to establish malicious intent. Thus the girls' confessions represented key evidence to establish intent and to prove arson.[7]

Extremely severe punishment, indeed. Yet in practice North Carolina did not execute for arson. By 1931, the state's electric chair, the preferred method of lethal punishment, had never seen an arson convict. As the Samarcand arson case pended, four capital offenses, homicide, rape, burglary, and arson, remained in North Carolina. Reform began before the Civil War, but it was the 1868 constitution that specified a dramatic reduction of the number of capital crimes, including prison breaking, slave stealing, horse stealing, inciting slave insurrections, bigamy, and any "crime against nature" (buggery, bestiality, and sodomy). Between 1910 and 1930, the state had executed 108 prisoners for the crimes of homicide, rape, and burglary. From 1910 to 1920 the state put to death thirty-six inmates for homicide, twelve for rape, and two for burglary. In the next decade, capital punishment for homicide increased by eight, while the number of executions for burglary and rape remained stable. Of the four death penalty offenses, the courts bestowed the greatest sympathy upon the state's pyromaniacs.[8]

The state showed leniency not only to pyromaniacs, but also to white female convicts. Between 1910 and 1930, North Carolina executed no women. Historically, capital executions of females were very rare, but they did occur. A 1989 landmark study of women and the death penalty has demonstrated that female offenders represent just 3 percent of those executed in America from 1608–1989. After the Civil War the number of female executions declined, as nine out of ten were put to death before 1866. Execution of juveniles is rarer still. In 1869, North Carolina entered the record books for executing one of the ten juvenile female offenders put to death in American history. The girl, Caroline Shipp, age seventeen and white, was convicted of murdering a white child. Historically, the U.S. South has led the nation with 60 percent of the cases of female executions, an incidence that peaked in the nineteenth century. Among death penalty offenders, the American South has executed 84 percent of the nation's African American women and 22 percent of its white women in 350 years.[9]

If numbers are any indication, then North Carolina reserved the death penalty for African American men. The U.S. South has developed a historical reputation for its brutal forms of violence and punishment, most notoriously in lynching and capital executions, to control African American populations. In North Carolina, the vast majority of executed inmates were African American men. Of the 108 executions from 1910 to 1930, the state executed just fourteen white men, twelve for homicide and two for rape. The precise relationship between capital executions and lynchings varied from state to state, but scholars agree that whites used both forms of killing to secure white supremacy. Lynchings declined in the 1930s as capital executions increased. However, state-authorized killings are not a substitute or surrogate ritual for mob violence. Particular cultural and historical processes shaped each lethal method. Lynching reinforced white men's patriarchal and social authority over women and African Americans. Lynch mobs typically targeted sex offenders and other deviants, while states largely authorized executions for homicides. The move from public hangings to executions performed behind prison doors suggests other changing cultural values. The spectacle and disorder that accompanied public executions, especially lynchings, undermined the efforts by southern lawmakers and authorities to promote restraint, discipline, control, and order in the form of private executions. Thus, white supremacy motivated both lynchings and state executions, but historical events shaped the cultural meanings as lynchings declined and state executions increased.[10]

A death penalty trial of sixteen girls in 1931 thus was coterminous to womanhood on trial. By historical measures, the threat of execution seemed unlikely for sixteen white female juvenile pyromaniacs. The state simply did not use execution to control this demographic group. North Carolina courts rarely sentenced whites, women, minors, or arsonists to death. To be sure, members of all these groups resided in the state's many public institutions, including county jails and poorhouses, delinquency institutions and insane asylums, and the state prison. Yet they generally were spared from death row. Under the aegis of white supremacy, white women's innate condition of superiority in terms of their chastity and obedience made it almost blasphemous to execute one even for the most heinous crime. She could be restored—unless, of course, she weren't really "white." As previous chapters have shown, "whiteness" did not apply to all the light-skinned. The wayward, the destitute, and the feeble-minded sometimes lost their "whiteness." Now would these girls die for the crime? Would the courts breathe sympathy upon rowdy delinquent juveniles or did the state intend to set an example by putting to death dangerous deviants?

The threat of execution was not real to some. It must have come as a great surprise to Agnes MacNaughton that she had elicited confessions to a charge with such a severe penalty. Apparently, most of the girls lacked any understanding of the seriousness of the case even after their arrests. Anyone paying attention on March 17, just five days after the fire, would have read that Magistrate George R. Humber had held a preliminary hearing that set a jury trial for the May term on charges of first-degree arson. The seriousness of the crime appeared to have minimal impact on the girls' actions in the ensuing weeks. Newspaper accounts reported that Margaret Pridgen, speaking freely to anyone who would listen, "takes entire responsibility for starting the fire." She fully confessed to lighting the fire when other girls failed in their attempts, and she claimed that she would "do it again" because she was "tired of the place" and thought she might be moved "if Chamberlain Hall burned to the ground." In fact, the girls said they preferred the Lumberton and Carthage jails to Samarcand and preferred to stay under the "fatherly guidance" of their jailer, Austin Smith. The *News and Observer* described their demeanor as: "Carefree and happy, most of them seem to be, little realizing the seriousness of the offense with which they are charged. Asked what punishment they expect, most of them reply, 'From three to five years in the pen.'" Perhaps the threat of death escaped them. Or possibly the girls suspected the state would not follow through on the threat precisely because they knew that the chief purpose of state executions was to control African American men. The girls may have presumed that their status as white protected them from extreme punishment. As runaways and mischief-makers, they had tested the boundaries of white womanhood, and it may have seemed to them that lighting matches and starting fires was hardly much more deviant than what landed them in the reformatory in the first place. Probably unaware of the controversial rhetoric of eugenics and the foreboding plight of the feebleminded, perhaps they misbehaved in accordance to their perceived understanding of themselves as white—immune to the harshest penalties of the law. It is also possible that the newspaper intended to portray the girls as capricious, wayward, and rebellious in order to maintain reader interest. However, the teenagers' actions, including Pridgen's testimonial, indicate that the girls possessed little to no fear of execution. That immunity would have been the privilege of whiteness.[11]

Youthful inexperience begets fearlessness, but so do anger and frustration. Most likely, the girls had developed a thick skin during one of the most dangerous, hostile, and deadly periods of North Carolina's history. Between 1929 and 1934, the Carolina Piedmont textile communities, home of some of the girls, had erupted into class warfare. North Carolina, the nation's largest textile state, shared with South Carolina an arc of mill villages inhabited by cotton mill people who abandoned farming in the Appalachian Mountains for work in the mill villages of the Piedmont. From 1905, textile workers lived in mill villages provided by mill owners and established their own communities and ways of life. After World War I, the demand for cotton cloth plummeted. The war boom had ended, hemlines had risen to the knee, and women no longer demanded cotton stockings. Faced with declining profits, mill owners responded with the "stretch-out," a term applied to the numerous strategies used to step up the pace of production. Owners replaced male operatives with women, increased the number of looms per operator, attached "hank clocks" or timing clocks to machinery, and restricted breaks and toilet visits. And wages stagnated, the average southern cotton textile wage remaining the same in 1927 as it had been in 1894.[12]

In 1929, mill operatives across the Carolina Piedmont struck in the tens of thousands. The walkouts persisted from March until October. On March 12, the entire workforce, more than three thousand mill operatives, walked out of the Bemberg and Glanzstoff rayon spinning plants in Elizabethton, Tennessee. Within weeks the strikes had spread to the Carolina Piedmont. Thousands of South Carolina workers in thirteen different mills responded to the stretch-out with walkouts. In North Carolina, operatives walked out of mills in Pineville, Forest City, Lexington, Bessemer City, Draper, and Charlotte. The conflicts that grew most deadly and gained the most national attention occurred in Gastonia at the Loray Mill and in Marion. The Loray Mill strike resulted in the shooting deaths of the police chief and strike activist Ella May Wiggins. In Marion, the governor turned out the National Guard when striking families resisted eviction orders from the mill villages in August. The Guard arrested 148 workers for various crimes, including "insurrection against the state of North Carolina." After another walkout two months later, the police killed six workers and wounded twenty-five.[13]

The hostilities of 1929 must have had some effect on the Samarcand defendants just two years later. Several of the girl defendants had lived or worked in these communities; two claimed to be children of cotton mill operatives and two had worked in the mills as well. Their stories about mill village life, walkouts, shootings, deaths, and the National Guard must have captured the attention of the girls from eastern North Carolina and likely provoked an awareness of and distaste for state authority. In fact, some conflicts were so new that they had not yet entered the region's collective memory. One very protracted strike at the Riverside and Dan River Cotton Mills Company in Danville, Virginia, the largest textile firm in the South, had ended in January 1931, just two months before the blazes at Samarcand.[14]

Surely the girls knew of the strikes, if for no other reason than that many of the strikers and their leaders were young women. Women represented 49.9 percent of the workforce in 1890 and 45 percent in 1900. One union representative in 1929 estimated that 60 percent of all textile workers were women. They were young. In 1920, the majority of women workers were between the ages of fifteen and twenty-four. They were militant. Young women workers had led the Elizabethton walkout, a conflict that ended only after the court issued an injunction against picketing and the National Guard had mounted machine guns on rooftops. Women's demands often centered on their roles as mothers. They sought shorter hours, higher wages, an end to stretch-outs and hank clocks, and better living conditions for their families. At the Loray Mill strike in Gastonia, many of the thirty strikers arrested were women. Mrs. Charles Corley was the first to be arrested, for wrestling a rifle from a Guardsman. She claimed that a "little thing like a bayonet would not stop her." National union organizations and the Communist Party sent women representatives to organize. The National Textile Workers Union sent Vera Buch and Ellen Dawson. Amy Schechter represented the Workers International Relief Organization.[15]

The Samarcand girls would have recognized the balladeer and most well-known woman leader, Ella May Wiggins, who, born in 1900, left her Appalachian family for the Piedmont textile mills. She bore nine children and, when her husband left her, supported her children by taking a job as a textile worker in the American Mills in Bessemer City. She was most famous for her ballads; often she sang about how the mills separated children from their mothers. Her song "Mill Mother's Lament" captured women's resentment at management for harsh separations and low wages:

> We leave our home in the morning
> We kiss our children goodbye
> While we slave for the bosses
> Our children scream and cry.
>
> And when we draw our money
> Our grocers' bills to pay
> Not a cent to keep for clothing
> Not a cent to lay away
>
> And on that very evening
> Our little one will say
> I need some shoes, dear mother
> And so does Sister May.

Now it grieves the heart of a mother
You every one must know
But we cannot buy for our children
Our wages are too low.

Now listen to the workers
Both women and you men
Let's win for them the victory
I'm sure t'will be no sin.[16]

As a union balladeer, Ella May Wiggins led a high-profile life, and it was this attention that likely led to her early and violent end. On September 14, 1929, Wiggins and other union members set out in a truck from Bessemer City for a union rally. As a result of aggressive or reckless driving of several cars following en route, their truck crashed. The other cars stopped and a mob gathered around the truck. Some fired their pistols at the truck. Wiggins was shot in the back and died almost instantly. She was twenty-nine. Word of her death quickly spread all over the world. The girls at Samarcand would have heard and discussed the sordid details. Nell Battle Lewis was so shocked that she could not still her own pen. She had sympathized with the striking workers as women and mothers of children. In her *News and Observer* column she decried the murder of a young mother who wanted "the possibility of a fuller life" for herself and her children. "The children . . . the baby . . ." Lewis pondered, "Dear Jesus what will happen to them now?"[17]

Years of angry protest and participation by young women provoked howls of disapproval from some southern whites, who condemned the women for acting against the submissive and demure characterizations of their gender and race. In press and in public, voices horrified by the behavior of the strikers cast the young women as unwomanly by portraying them as upending the social order and as being dangerous anarchists, communists, and advocates of sexual freedom. The sharpest critics compared the strike to the Bolshevik Revolution. The *Gastonia (N.C.) Daily Gazette* described the strike leadership as "against all American traditions and American government!" The *Gazette* predicted a reign of terror unless the state suppressed the strikes. "Whenever the Communists get their bloody claws on America anarchy will reign and there will be no mills or factories to work in." The Loray Mill strike, the *Gazette* cried, "WAS STARTED FOR THE PURPOSE OF OVERTHROWING THIS GOVERNMENT AND DESTROYING PROPERTY AND TO KILL, KILL, KILL!" The *Southern Textile Bulletin,* a textile industry trade journal, argued that communists had provoked the strike to spread sexual promiscuity and interracial intermixing. The "doctrine of full love, no religion, and social equality with the negroes" were central to communist beliefs, argued editor David Clark.[18]

To the media, young women strikers seemed a manifestation of these supposed evils. By engaging in the strikes, they had refused to embrace their destiny as southern ladies and the keepers of the status quo. Young women textile strikers had embraced the provocative 1920s youth culture with their own working-class flair. Fast music, radios, cars, and a dating culture epitomized changing values nationwide. In the Carolina Piedmont, young women strikers adopted forms of dress, language, and behavior that many observers deemed disorderly. Women gave fiery speeches at rallies. Others wore overalls and men's clothing while picketing. A new youth culture arose as a form of working-class women's protest that subverted gender norms. Critics of the day said it was simply inappropriate for respectable ladies to picket in the streets. "It isn't decent," read one letter to the *Gazette,* "I have seen young girls, I mean strikers, going up and down the street with old overalls on and men's caps, with the bills turned behind, cursing and calling the cops all kind of dirty things."[19]

Only bad girls taunted the police. The media had employed the same imagery just two years earlier to characterize the 1929 textile strikers, and now they used them again with the 1931 arson defendants. As a witness to both conflicts, Nell Battle Lewis found she could not escape the similarities in public sentiments. To muster any able defense, Lewis knew she must address head-on public angst over the behavior of poor, white, female adolescents. The 1920s and 1930s South was grounded in a racial caste system that stressed the pristine virtues of the southern lady. Many southern whites viewed as disorderly any poor white girls who failed to meet the standards of racial and sexual purity. Courts, politicians, and the media associated their bodies with disease, sexual vice, crime, and feeblemindedness. Keenly aware of her clients' intrinsic image problem, Nell Battle Lewis knew that this case represented southerners' sexual anxieties as much as it did the crime of arson. The legal and technical details of the case rested on the crime of arson. However, the trial and swirl of public controversy repeatedly turned to the issues of sexuality, womanhood, and the sexual delinquency and mental state of the girls to explain their riotous actions. Rebellious sexuality is how the *Moore County News* described the girls involved in smashing windows at the jail. In unjournalistic prose, the reporter described the defendants thus: "Their faces, mostly pretty, distorted with rage; clothing disheveled, hair awry, and eyes gleaming, they seemed to be angered to the point of temporary insanity. Strangely, though, they have failed to divulge any immediate reason for their revolt." Little in the way of humanity marked this journalist's prose. Pretty but enraged by insanity, these girls were devoid of reason and, like rabid animals, on the prowl.[20]

Nell Battle Lewis fully understood that any successful defense of her clients required a powerful counterargument to the image of unbridled sexuality and rebelliousness unleashed by the press. Her trial preparation, including her investigation, selection, and questioning of witnesses and opening and closing remarks all reflected a consistent strategy to present her clients as neglected children who were failed not only by their parents but also by the state. Victims of their environment, the children were products of social forces beyond their control. She drew her argument from prevailing studies of the psychology of the age. The psychology of adolescence in 1931 was heavily indebted to the turn-of-the-century psychologist G. Stanley Hall, founder of the American Psychological Association and the *Journal of Psychology*. In his 1905 tome, *Adolescence,* Hall argued that the teenage years was a stage of development unique to the modern industrial era. Adolescence marked a time of "marvelous new birth" when a child transitioned from a barbarous state to a civilized one. In adolescence, children acquired muscular development, characteristics of reasoning, fine motor skills, and sexual, moral, and social sensibilities. As such, adolescence marked a period of great danger, as a bad environment could lead adolescents astray.[21]

Hall claimed that adolescent development differed by sex. Girls, he claimed, remained perennially adolescent. Unlike boys, who could attain a higher state of civilization, girls could never attain a full state of rationality because of the monthly interruption of the menstrual cycle. Nonetheless, if properly cultivated, a woman's perpetual adolescence possessed many charms. "Woman at her best never outgrows adolescence as man does," Hall argued, "but lingers in, magnifies and glorifies this culminating stage of life with its all-sided interests, its convertibility of emotions, its enthusiasm and zest for all that is good, beautiful, true and heroic. This constitutes her freshness and charm, even in age, and makes her by nature more humanistic than man, more sympathetic and appreciative." By 1928, feminist psychologists had modified Hall's gender pro-scriptions to claim that girls could attain those previously ascribed masculine traits. Feminists, including Leta Stetter-Hollingsworth and Phyllis Blanchard, shifted the focus to argue that adolescent girls, in their adjustments to modern society, represented a sympathetic figure whose behavior should be explained and adjusted. Episodic disorderly behaviors were normal to adolescent devel-opment, they argued. "We have been having an interlude of women as peren-nial adolescents," psychologist Lorine Prudette ruefully remarked, "but there are signs that the interlude is about over."[22]

The path to womanhood, according to the psychological wisdom of the 1920s, thus allowed girls a period of rebellion, known as adolescence. While the experts differed as to the degree of rationality attainable by adult women, they agreed that adolescence provided teenagers with a period of life-learning lessons that allowed for individual testing of social boundaries, experimenta-tion, and petty delinquency. Nell Battle Lewis understood that this philosophy of adolescence might have a national audience. The Samarcand arson case pre-sented her the challenge of convincing a southern white audience, steeped in the conventions of gender and race, that this formative development process might exist among southern white girls as well.

It seemed the prevailing literature gave Lewis the upper hand, except for one very disturbing detail. Just two days after the second jail fire, the North Carolina General Assembly had passed a joint resolution that effectively stripped the girls of their juvenile status. Senators from Moore and Duplin Counties had introduced the resolution with the intent to bypass the State Child Welfare Act, legislation that afforded protection to minors by establish-ing juvenile courts and juvenile reform facilities so they would not be tried or imprisoned as adults. The resolution, "Relative to the Safe Keeping of the Girls, Formerly Inmates of Samarcand, Now Under Indictment for Firing the Building of that Institution," legally nullified the girls' juvenile status and al-lowed, at the governor's discretion, their imprisonment in any facility in the state. Several senators objected to the resolution for the reason that to house the girls in the state prison, itself a "fire trap," would present a danger to the hundreds of adult convicts there. Despite these protests, the bill passed. No longer considered by the state to be children, the girls would face prosecution, incarceration, and possibly execution as adults.[23]

Lewis used her connections with the press to strike back. She provided information to sympathetic journalists and allowed them access to her clients for interviews. On May 2, the same day that the *News and Observer* reported passage of the incendiary bill, the newspaper also published a lengthy piece, "Rebellious Girls Set Their Bunks Afire to Get Thrills," by reporter Carolyn L. Reynolds. Reynolds's headlines and text described the girls' behavior as rebellious, disturbing, and disorderly, but interpreted their actions as the kind of behavior expected from adolescent girls long deprived of positive home environments. Like the feminist psychologists who argued that girls exhibited a wide range of behaviors depending on their circumstances at home, Reynolds suggested that society had gone wrong for these girls. Bad home environments, neglect from the state, and long periods of confinement had stunted the girls' adolescent development. "I just feel mean," Reynolds quoted fourteen-year-old Margaret Pridgen as saying when asked why she participated in the outbursts. Even the subhead suggested that confinement had made the girls loopy. "Samarcand Pupils Sick of Confinement in Waiting for Trial," "Highly Emotional, Crave Excitement," and "Must Do Something to Break Monotony of Darkened Jail Cells" all prefaced the articles to explain how these normal little girls, she said, had "gone a bit haywire." Reynolds strongly asserted that the girls had acted out not from mistreatment but from boredom and restlessness, common traits that beset adolescents when deprived of play.[24]

Her article included numerous insights into the girls' potential rationality. "Six of the girls when interviewed in Moore County jail yesterday," she reported, "were unanimous in their agreement that they have not been mistreated in any way while in the jail. In fact, they insist that Sheriff McDonald and his deputies have been very kind and are 'mighty nice.'" Other girls, Reynolds reported, were eager to set the record straight. Sixteen-year-old Josephine French objected to the account of their behavior as a "rampage." "I wonder who told them that," Reynolds reported from her interview with French. French explained that the girls set the fire when the warden's wife forgot a promise to allow the girls some time to play.

> I'd like to get the thing in the paper just like it happened. We were in the room (she referred to the cell) and asked the jailor's wife to let us out for a while. They do let us out of here every day. There's a victrola upstairs and we can go and play it sometimes and dance. She said she couldn't. We wanted to get out

and she said she'd see about it and went downstairs. She didn't come back so we set fire to one bunk—to bring somebody up to let us out. We supposed they had forgotten. They came then—and let us out. Down there, some went one way and some another.[25]

The police overreacted, the girls claimed in the interview. The fire hose soaked them in the process of putting out the fire and no one stabbed the barber in the arm with a knife. According to Reynolds, the girls explained to her in "shrieks of girlish excitement" that "he got cut jumping out the window. There were only two girls had knives and they wouldn't cut. We didn't try to fight anybody. We just wanted to get out of this room for awhile." The jailor and sheriff said they simply had found themselves overwhelmed by the management of these "high-strung girls."[26]

If not for their unfortunate circumstances, Reynolds declared, the girls would be "average high school girls." To gain reader sympathy, Reynolds drew upon popular ideas of adolescence to depict the girls as unfortunate victims of their society. Reynolds reported that they wore "clean, brightly colored prints dresses, stockings rolled to the ankles with one exception, overflowing with energy, all trying to talk at once." The girls were no different from any other high school-age groups. Like any other girls, they ostracized those who were different. In jail they targeted fourteen-year-old Margaret Pridgen, who lived in a separate cell and had not participated in the latest rebellions. They teased her and berated her. Reynolds, who had come to sympathize and relate to the girls she interviewed, began to internalize the girls' criticism of Margaret. In the article, Reynolds described Margaret's adolescent development as a failure in the extreme and implicated her as being complicit in the original arson case. While the other girls appeared normal, Margaret, Reynolds explained, was ruined. "Bitter" and "cowed" by her father's abuse and her long confinement at Samarcand, she no longer appeared normal but "beaten by the circumstances that have overtaken her." The experience had badly destroyed her personality and even her physical features, Reynolds noted with despair. "Her mouth droops" and her "liquid brown eyes" are "longing with hopelessness." Pridgen's lost soul, Reynolds suggested, explained the horrible crime, while the other "fresh-faced" girls probably were blameless.[27]

This journalistic characterization of Pridgen endangered the case. Reynolds's comments echoed her contemporaries in similar characterizations of the mentally unfit poor. The *Moore County News* had used similar prose to describe the girls as possessing faces "distorted with rage," "eyes gleaming," and "angered to the point of temporary insanity." Kate Burr Johnson had used animalistic descriptions of behavior and features to describe the Fehler family, her example in the professional literature of the worst extent of genetic mutation in a family line. Red Fehler, she had written, howled like a wolf and climbed trees. These North Carolina journalists likely had picked up the language and style from the well-acclaimed work of a colleague in Georgia. Erskine Caldwell had written several best-selling depression-themed novels in the early 1930s that described the animalistic characterizations and feebleminded personalities which had infected families mired in long-term poverty. Reynolds had spared fifteen of the girls from this journalistic trope. Instead, she had endangered the strategy of her friend, Nell Battle Lewis, to defend all of the girls as adolescents by her portrait of Margaret Pridgen as mentally insane. Evidence suggests that several girls were culpable, but Reynolds's account pinned the blame all on one presumably mentally defective girl.[28]

A week later, the girls set yet a third jailhouse fire. This time Nell Battle Lewis was on the case. Her own careful handling of the reporters likely influenced the sympathetic article that appeared in the *Moore County News*. The very subdued headline did not mention the blaze at all. "Samarcand Girls Quietly Awaiting News of Their Fate," announced the newspaper, and in smaller type: "small blaze started by one girl proved of no consequence." Only one girl appeared at fault, and all the others apparently bestirred themselves to extinguish it. Even "little Margaret Pridgen" seemed "done with any further thought of insurrection or rebellion." Mindful of the sensationalistic potential of this last fire, this journalist cautiously noted the split public opinion on the girls' fates. "Many [people]," the article noted, "are outspoken in defense of the county's young charges; others believed to be in the minority, are avowedly of the belief that drastic measures should be taken in the disciplinary punishment to be accorded them." Once again, Lewis proved she had some control over the media in reporting news on her defendants.[29]

Hope springs eternal. Lewis had reason to believe she could win the case. A Cabarrus County judge had ruled leniently in a male juvenile case, that of Edward Eggers, age fifteen, who had set fire to a cottage at the state's juvenile reformatory for delinquent white males, the Stonewall Jackson Training Institute. To a packed courtroom, Eggers had confessed, adding that he was whipped until he implicated two others in the crime. School officials defended the whippings, and one board member argued that if these bad boys "were sent back to the school there wouldn't be a building standing in six months." Counsel for the defense charged that the confessions followed severe beatings at the hands of the instructors. Evidently, the judge also had found the abuse unwarranted and ruled leniently toward the boy. [30]

On May 19, 1931, Lewis unwaveringly and steadfastly assumed the same defense. It was not always easy to promote her defendants as innocent children. The press wanted to sensationalize the bad girls as much as ever. The girls ranged in age from thirteen to seventeen years old, and for all intents and purposes they were beyond any stage of childhood. Newspaper photos depicted young women ascending the courthouse steps on the morning of the trial. Lewis had clad them "in attractive silk and cotton prints" and had transported them to the courthouse in a school bus. The girls exited the bus and crossed the street, where a reporter for the *News and Observer* waited to capture a photo. He snapped a picture just at the moment when one of the girls drew heavily upon a cigarette and lifted a leg to step up to the curb. The unflattering photograph and caption that appeared in print the next day played upon the public's impression of working-class girls being sexually rebellious, and the exposed leg and indignant cigarette smoking left little to the imagination. [31]

Nonetheless, Lewis persisted. Even before the opening gavel, she had secured a plea bargain of guilt to the lesser offense of attempt to commit arson, a felony punishable by four months to ten years imprisonment. It was a big break that occurred in pretrial negotiations with the solicitor. The state would not execute the girls, but the trial would go on. The reduced offense still required a sentencing trial in an adult court. The judge might dismiss any defendant for lack of evidence, an agreement also sufficient to the solicitor. Upon his opening remarks he announced that he did not expect that public opinion would demand the state "try the young girls for their lives." Almost immediately he dropped charges against two girls, Mary Bronson of Rocky Mount and Wilma Owens of Waynesville, for insufficient evidence. The death penalty averted, now the trial proceeded to determine which girls were guilty and what sort of punishment each might suffer for the crime. [32]

The trial commenced. The prosecution, led by State Solicitor Don Phillips, presented its case in the morning. The state's case rested on evidence of the girls' alleged confessions. As his first witness, Phillips called Estelle Stott, secretary to the superintendent, who testified that twelve girls had confessed to "actually setting the building on fire" that very night. The other girls confessed or were implicated by others on the next day. Lillian Crenshaw also took the stand for the prosecution. Phillips introduced Crenshaw as being the supervisor of "student government," though Lewis had identified her in a list of witnesses with a more nefarious-sounding title: "Discipline Officer." Crenshaw, who also was present at the alleged confessions, verified Stott's testimony. The girls, both women stated, had confessed to setting the buildings aflame. Not called to the stand was Agnes MacNaughton, the superintendent herself. Perhaps Phillips believed that the headmistress's caustic personality might not sit well with the judge.[33]

On cross-examination, Lewis interrogated the "Discipline Officer." Counsel intended to prove that the state, as *parens patriae*, not only had committed neglect in its role as guardian of the children, but also had fostered a hostile atmosphere in the institution that furthered the girls' rebellion. With Crenshaw on the stand, Lewis immediately questioned her to reveal the practice of corporal punishment. How were the girls disciplined? The "Discipline Officer," still under oath, played right into her hands. Crenshaw admitted to the whippings. What were the circumstances of the punishments and who supervised? Typically, MacNaughton forced the offending student to lie down on a rug while she or another faculty administered the whippings. Had Crenshaw witnessed any of the whippings? She had, but none were performed unmercifully. Thank you, you may step down. The judge broke for lunch.[34]

The morning cross had positioned Lewis for her afternoon defense. When the trial resumed, Lewis called to the stand seven of the girls, who proceeded to testify to having experienced brutal rounds of whippings. At least ten of the fourteen defendants claimed to have received whippings, some more than once, since entering the institution. Girls described the implements used as "big switches," "sticks," or razor "strops." Some claimed they endured one hundred lashes at once. Josephine French, committed to Samarcand in 1930 for "running around," testified to one whipping at Samarcand, as did Dolores Seawell, who added that her hair had once been cut off for punishment. Edna Clark, committed for "being a run away with bad associates," testified to bearing three whippings. Margaret Abernethy swore under oath that she had received four whippings with straps and switches, "twice for running away and twice for 'being rude to a teacher.'" Abernethy and other girls testified that they were locked in rooms for as long as three months. "The room," she said "was so filled with bed bugs that sleep was next to impossible." Bugs were bad all the time, she said, but "worse when one was locked in the room all the time." As sheets were not allowed in locked rooms, "girls slept on blankets placed on the bed or floor."[35]

Would Judge Michael Schenck believe what he had just heard? In case he did not, Lewis called two former high school teachers, Lottie Mitchem of Raleigh and Fronie Harrell of Durham. MacNaughton had hired Mitchem on September 11, 1930, and had dismissed her on February 28, 1931, shortly before the fire, for causing a "disturbing influence." In a deposition to Lewis, Mitchem claimed to have left on her own accord in protest against MacNaughton's orders that she falsify school attendance records to show that girls were attending classes when, in reality, they were locked up. Mitchem and Harrell confirmed the girls' testimonies of bed bugs and unsanitary conditions. While neither had witnessed any whippings, Mitchem stated that she had heard the girls' cries and knew that some had to seek hospital treatment afterward. "I heard the girls screaming several times from my building (Bickett)," she told Lewis when deposed, "and once I heard them screaming when I was in the school building." Both women further testified that girls with venereal diseases were confined with noninfected girls and were required to use the same bathroom facilities.[36]

Throughout her cross-examination and her defense, Nell Battle Lewis stuck to her strategy. The court should dismiss or lightly sentence the girls because they were innocent adolescents and victims of state neglect. The authorities had punished the girls brutally and inhumanely. But were the girls guilty of arson? Did their victimization absolve them of their guilt? All seven of the defendants who testified claimed that they had not made any confession to Estelle Stott, or that they had done so spuriously in order to escape the institution. In other words, the girls themselves professed innocence. Except for the initial confessions on the night of the fire, nearly all of the girls consistently professed their innocence. Margaret Pridgen, who did not testify, was the only girl who admitted guilt while in jail. During incarceration, Pearl Stiles maintained her innocence in a letter she composed to Governor O. Max Gardner in which she pleaded for mercy for herself and the other girls. "The way we were treated was terrible," Stiles explained. "We were locked, beat, and fed on bread and water most of the time. Please give me liberty or death. . . . We girls in Robeson County jail is just as innocent of this crime they hold against us as a little child. Mr. Gardner, this is Pearl Stiles writing and I am always trying to be good." Apparently the letter never reached the governor but was intercepted and filed with other records of the State Board of Charities and Public Welfare. Even Margaret Abernethy professed innocence and that her confession was made under duress. Abernethy testified that she had confessed to helping set the fire, but she did so in hopes that she would be sent to jail, where conditions and treatment would be better than what she had endured at Samarcand. If the confessions were made under duress and the girls were proved innocent, then the judge should sentence lightly.[37]

The solicitor had accepted a plea to the lesser offense of attempt to commit arson, and the judge was to determine the length of sentence, from four months to ten years imprisonment. Lewis intended to show the judge that the girls were mere children and not fully culpable. As children they merited leniency. The children were wards of the state, Lewis argued, and North Carolina was responsible for their welfare. The state had recognized their problems and should have disciplined them fairly and reasonably. "I am not saying they should not be punished," Lewis argued, "but I am saying that half-grown girls in a civilized community should not be laid on a whipping carpet, when flogging has been abolished in chain-gangs."[38]

Then, in the last minutes of her defense, Nell Battle Lewis shifted course. Forever hopeful that she might make an argument to lessen her defendants' punishment, she employed an insanity defense. She had never floated this argument with the press, possibly because she feared a backlash from the public. The shift probably occurred because she had some experience as an advocate of the insane. As a reformer and advocate of the downtrodden, she had served as an advocate of the mentally unfit. In her *News and Observer* column, "Incidentally," Lewis had posed questions about the rights of the mentally ill. The mentally ill, she believed, did not bear the same culpability for crimes as normal citizens because they did not understand the crime they had committed. She advocated preventative treatment, reformatories for the insane, and sterilization. Her critiques of capital punishment and the brutalities of the electric chair rested fundamentally on her concerns about mental health and the rights of the insane.[39]

Perhaps her agenda as a progressive reformer caused her to take the one extra step to argue that her defendants were feebleminded. It really was not necessary, as she already had secured a lesser sentence. Also, it had not been her plan as part of her media campaign before the trial. Perhaps she employed the shift due to a lack of confidence. In her first case, she might not have seen the forest for the trees. She was, in fact, in control. The case had a chance of dismissal or significantly reduced sentences for some or all of the girls. Nonetheless, she entered the insanity plea at the end of the trial. As her last witness, Lewis called to the stand Dr. Harry W. Crane, a professor of psychology at the University of North Carolina and the director of the Bureau of Mental Health and Hygiene of the State Board of Charities and Public Welfare. Crane testified that he had examined four of the defendants and that their mental ages ranged from eight years, ten months to eleven years, one month. In his opinion, he added, the girls were feebleminded and unable to differentiate between right and wrong. Without a matured sense of situational morality, Crane argued, the children "would only know that certain things had been indicated to them as right or wrong." As Crane's testimony indicates, the case did not rest on the legality of the purported confessions but on the girls' mental states at the time of the fire. "These children are all young, with their eyes to the future," Lewis pleaded in her closing, "and I think the state whose wards they were when the crime was committed should give them as great a chance as possible." Judge Schenck must exhibit sympathy to the girls, Lewis concluded, for as victims of their environments their mental capacities had not matured beyond the state of childhood.[40]

Judge Schenck was not persuaded. It is possible he found the shift in strategy annoying and unprofessional. He accepted neither the first defense, that the girls were victims of state neglect, nor the second strategy, the insanity defense. He said the girls, though adolescents and rebellious, were of sound mind. Sentencing came the next day. Twelve of the girls received five-year sentences in the state penitentiary, to be released in eighteen months if they "behaved," he said. He granted suspended sentences to Rosa Mull, age thirteen, and Margaret Pridgen, age fifteen. Schenck promised Mull, the youngest of the defendants, custody with her parents and a five-year suspended sentence that might be shortened to two years on good behavior. Pridgen's parents were dead, so she would live with relatives or in another home, but would receive a suspended sentence of one to three years. Schenck did not specify the reason why Pridgen received one of the lightest sentences. However, Pridgen, the only penitent defendant, was the only girl consistently portrayed by the media as mentally defective. Perhaps the insanity plea was a factor that worked in her favor after all. Schenck did not elaborate.[41]

The other twelve would immediately report to the state penitentiary in Raleigh, where they would be housed above "Death Row," as it was the only fireproof wing in the prison. Neither Lewis's closing remarks nor Crane's expert testimony had persuaded Schenck that the twelve girls possessed the mentalities of children or mental defectives. He spoke to them as if they were adults. As he recounted their offenses, he told them he thought they could understand him. He had, Schenck argued, exhibited undue leniency. The girls had been indicted for grand arson, and could have been tried for their lives. He dismissed all accusations of neglect against the institution. These girls were sent to Samarcand because they could not be controlled at home. The state "had done the best it could for them at Samarcand," he said, but they remained unruly and rebellious. Only the state penitentiary could safely hold them, he maintained. As if echoing the resolution by the General Assembly, he argued that the firebugs threatened the safety of other inmates unless placed in a fireproof facility.[42]

One might say that Nell Battle Lewis's strategy of portraying the girls as adolescent and then as feebleminded victims of unfortunate circumstances had failed. In a society that accepted the concept of adolescence, the court had issued to twelve juveniles a weighty sentence reserved only for adults. However, observers exhibited sympathy. The *News and Observer* quoted several unnamed spectators to prove the prevailing sentiment that the girls had received too heavy a sentence. "It does look like what they need is help, not punishment," observed one courthouse employee. A service attendant at a gas station agreed. "Them girls have already been punished twice over for everything they've done. They've been punished ever since they were born." Another spectator observed, "it is a case of being more sinned against than sinning." The unfair sentence, the *News and Observer* concluded, "will do little toward improving the girls as citizens of the state."[43]

Why did Judge Schenck disregard the arguments of the defense? It is likely the answer is found not in the courtroom but in the historical context in which the trial occurred. Historically, North Carolina treated white women, youth, and arson defendants with some leniency and reserved the worst punishments for African American men. However, 1929–31 was a tumultuous period in North Carolina. Textile mill owners and employees engaged in mortal combat and young women employees often positioned themselves at the heart of the conflict. Some members of society thus distrusted women of the working class and remained suspect of their sexuality. Judge Schenck may have ruled in this context. On her part, Nell Battle Lewis did not provide the iron-clad defense that might have turned the tide. In her quixotic defense, Lewis first represented her clients as unfortunate adolescents en route to southern white womanhood, yet victims of state neglect. In the end, she adopted a new course that depicted the girls as feebleminded and mentally defective. She likely did so as an advocate for the mentally insane, though any characterization of the girls as mentally defective risked playing into the hands of the girls' worst critics. Regardless of her intentions, this legal detour compromised her clients' claims to the psychology of adolescence, disrupted her courtroom momentum, and possibly angered the judge. Twelve of her clients received a maximum sentence of five years above Death Row in the state penitentiary.

Yet the story about the Samarcand Arson case does not end with the sentencing. Nell Battle Lewis's public strategy to spark sympathy for the girls had some effect. Numerous southern voices expressed sympathy for the Samarcand girls. Not only their lawyer, but sometimes the media and local spectators argued in their defense. Even the state solicitor had agreed to a plea bargain that dropped the capital punishment charge. What really happened that night of the fire? Evidence and testimony suggest that probably Margaret Pridgen, Margaret Abernethy, and Marion Mercer had conspired in setting the fires. There is no evidence that the other girls played a direct role in the crime. Several more confessed to the jail fires set in protest, they claimed, against bad conditions. Some of the girls were guilty. Significantly, the trial had sparked the minds of the public. In the days immediately following the trial, an uprising of progressive voices from across the state criticized the trial's outcome and demanded a state investigation of the most offensive conditions at Samarcand.

7

Classifying "Subnormal"

PAROLE AND STERILIZATION AT THE STATE HOME AND INDUSTRIAL SCHOOL FOR GIRLS, 1933–1950

IN MAY 1931, TWELVE TEENAGE ARSON CONVICTS found their new home among a population of adult women in the state's Central Prison. The state seemed to have solved its immediate problems in incarcerating the Samarcand firebugs. But what did authorities intend for the remaining unruly white adolescent girls in the state? This chapter explores the controversy surrounding Samarcand and its inmates after the 1931 trial. In the months after the trial, reformers, parents, teachers, and state authorities argued about the governance of unruly white girls. On the one hand, some residents continued to presume the inherent redeemable qualities in white girl inmates and demanded investigation and reform at Samarcand. Others viewed this class of white girls as mentally unfit and unreformable because their condition as mental defectives separated them from the more pure lines in the race. By 1933, the authorities governing the state's juvenile delinquency program had abandoned the reform principle that had previously governed discipline and training at Samarcand Manor. The Board of Managers instead adopted new white purity policies based on eugenic theory and racial hygiene laws practiced in Nazi Germany. The Samarcand board and the superintendent institutionalized a classification and parole system intended to identify mentally unfit girls for the purposes of sterilizing them before their reintroduction into society. From 1929 to 1950, half of all state sterilizations were performed on girls aged ten to nineteen, and 12 percent (293 of 2538) were performed on juveniles at Samarcand.

The Samarcand arson case and investigation served as a turning point in North Carolina's public policy history. This public policy shift, from reform and redemption to classification and parole, represented a new construction of white supremacy, a racism that defined whiteness more narrowly and stripped its privileges from light-skinned girls on the lowest economic rung. The state investigation at Samarcand engaged many North Carolinians in a conversation about whiteness and a discussion about who enjoyed its privileges. Many parents, teachers, state authorities, and members of the press maintained their faith in the potential of the state's future southern ladies at Samarcand. Extreme forms of punishment and negligent conditions, they argued, should not be tolerated at the institution. Though economically disadvantaged, these girls at the bottom rung could still be saved. Yet the mental hygienists were gaining steam politically. The 1931 trial had introduced very negative perceptions in the public eye of Samarcand girls as unredeemable and feebleminded delinquents. Sterilization laws, at first weakly enforced, strengthened throughout the investigation and trial. In 1933, as Nazi Germany embarked upon its sterilization campaign, North Carolina's General Assembly passed its own sterilization law. The 1933 law was the state's third attempt, but it represented the version most resilient to legal challenges, and it passed simultaneous with the conclusion of the Samarcand investigation and the replacement of Superintendent Agnes MacNaughton with a known hygienist, Grace M. Robson. This chapter argues that the events of 1933—the Nazi sterilization campaign, North Carolina's revised sterilization law, the Samarcand investigation, and the subsequent appointment of Grace M. Robson as Samarcand Manor superintendent—were not coincidences. All of these events that year marked a significant shift in a definition of whiteness rooted in the exclusion of a class; this whiteness demonized the poor as no longer white, declared them unfit to reproduce, and fostered a decades-long repressive sterilization scheme to control poor white women and girls no longer considered southern ladies.

Before 1933, no one seriously argued that Samarcand girls should be sterilized. Sterilization had its advocates in the state and nation, but the procedure is never mentioned in the Samarcand records of the arson case or in the records of the subsequent state investigation that took place at Samarcand Manor during the spring and summer of 1931. The fire, the May 18 trial, and the state investigation sparked widespread outrage among citizens, newspapers, teachers, former teachers, and local and state public welfare officials. Most of the public indignation revolved around the matter of corporal punishment, teacher turnover, and the question of segregation and treatment of girls infected with venereal disease. Reports of abuses and complaints directed attacks at Superintendent Agnes B. MacNaughton, the aging reformer who had always envisioned Samarcand as a school where white girls could restore their tarnished reputations. MacNaughton survived the attacks for several months, but eventually the board replaced her with an up-and-coming mental hygienist who championed a new parole and classification system that included sterilization.

After the 1931 trial, North Carolinians demanded disciplinary reform, not sterilization. In the days and weeks after the fire and again after the May 18 trial, letters from parents and teachers and newspaper editorials expressed outrage at the conditions girls were forced to endure. Authorities had committed untold abuses on the girls of Samarcand. Media coverage of the post-trial investigations suggests a public in high anticipation of action. Reporters, editors, and concerned citizens demanded an investigation. The *Chapel Hill Weekly* angrily compared Samarcand to a "callous" penitentiary, though it was created to rescue and redeem girls of tender years. Other newspapers, such as the *Rocky Mount Telegram*, condemned the riots at Samarcand as a "serious indictment of the school," where the board of trustees and officials of the school were "guilty of gross neglect." Conditions had to be dire, concluded the *Chapel Hill Weekly*. "Something made these children come to the point of revolt and frenzy. It must have been extraordinary to have raised such a spirit of rebellion in such a type of girls." The state, it concluded, "must make the conditions known to the public and correct the problems."[1] Other articles demanding reform and investigation appeared in the *Greensboro News* and the *High Point Enterprise*.

The attacks also came from outside the state. Former teachers who had relocated elsewhere leveled serious charges on Agnes B. MacNaughton and the administration at Samarcand. The written complaints found their way into the hands of Lillie M. Mebane, legislator and chairwoman of the Public Welfare Committee in the North Carolina House of Representatives. Mebane pressured Dr. Delia Dixon-Carroll of the Board of Charities and Public Welfare, who demanded action from Public Welfare Commissioner Annie Kizer Bost. In April 1931, three weeks after the fire, Commissioner Bost assigned three staff members of the State Board of Charities and Public Welfare to investigate conditions at Samarcand Manor and to interview former and current employees who might shed light on the circumstances of the fire.[2]

Former staff, including a school nurse and teachers, reported terrible conditions and cruel disciplinary procedures in January and February 1931, just prior to the fire. Former nurse Viola Sistare left Samarcand on February 18, 1931, two weeks before the fire. Sistare complained that the administration had not afforded her the professional autonomy and respect due to her as a trained nurse. "Their commands were of little or no sense," Sistare argued, and "if followed would have resulted most disastrously for the three hundred children under my care." The staff refused to allow Sistare to provide proper medical care. In the hospital, accommodations were crowded and sanitation was poor. Girls infected with gonorrhea and syphilis received good treatment, but they shared facilities—including bathtubs, toilets, and even dishes and clothing—with uninfected girls. Sistare reported cruel whippings of girls by administrators and argued that staff sometimes intervened in her treatment of abused girls. Other teachers confirmed Sistare's complaints. Arline James, Georgia Piland, and former high school principal Lottie Mitchem reported unsanitary conditions, corporal punishment, and falsifications of school attendance records. According to several former teachers, the administration falsified monthly school records to show Moore County school officials that students were marked as present when some were in the hospital or "locked up."[3]

Sistare and the other teachers had departed Samarcand in late February, as part of a mass exodus of teachers and staff in the middle of the 1930–31 school year. Teacher turnover was very high, an indication of low pay and low morale. In March 1931, Samarcand employed thirty-seven teachers and staff. Ten teachers had departed in the first three months of 1931; ninety-six had left since 1929. According to staff records, many left for work elsewhere, or had fallen ill, or were employed for "summer work." But of the ten departures of 1931, five were discharged as "not suitable" or "unsatisfactory," or for creating a "disturbing influence." Former staff expressed great animosity toward the administration, especially toward Mrs. MacNaughton. Former principal Grace Henslee not only confirmed accounts of whippings, but also complained of the "unhappy frame of mind of the workers." No one knew in the morning, she stated, if they were to be fired by nightfall. "The workers spend a large percentage of the time they could be doing good locked in their rooms crying," she reported.[4]

Perhaps the most damning of accusations came from parents and teachers who told stories about abusive whippings. Several teachers reported knowledge of whippings, though all said they had never witnessed one. Georgia Piland complained that she had heard screams and was told that hickory sticks were used to beat the girls. Two girls had told Lottie Mitchem that they suffered as many as three hundred strokes given by as many as four people at one time. Nurse Sistare reported that she was called to provide a "pill" to quiet a girl after a whipping. Girls, at the risk of further punishment, reported their horrors to their parents. Most parents worried that to report these complaints would further harm their daughters. One mother was granted only a fifteen-minute visit with her daughter, confined in the discipline dorm. In that short visit she learned about the severe beatings that girls endured at the hands of the staff and other girls. She drafted these complaints in a letter, signed "Citizen," to the *News and Observer* in protest. She entitled the letter "Smoke Signals from Samarcand," referencing the arson case as a plea for help against a brutal and arbitrary administration. Girls were whipped with "bunches of five basket reeds" as many as fifty-one times, "Citizen" argued. "Some of the girls," she said, "have not been able to lie on their backs for several days after they are punished." It was customary for girls to whip runaways, supervised by staff. When one girl, Mary Horton, refused to take part in whipping a friend, she in turn suffered as punishment "one of the worst whippings ever witnessed." "Citizen" described beatings of very young girls. Elsie Louis Clinard (five to seven years old) was whipped "over and over for bed wetting." Two girls, Mary Ellington and Delia Frazier, told their parents about suicide attempts because of confinement and whippings they received after running away. Cruel whippings involving girls beating their friends created an atmosphere of terror in the place. "It's a wonder," Citizen noted, "this institution hasn't burned long since from spontaneous combustion during one of these switchings." It is unclear whether the *News and Observer* ever printed the letter, but a letter that contained similar content and also signed "Citizen" appeared on the editorial page of the May 4, 1931, *Fayetteville Observer*.[5]

Most staff members condoned the whippings. Staff who had witnessed whippings reported that girls lay fully dressed on a rug while a staff member administered six to ten licks with a "small light hickory switch." Investigators found that most staff members approved of corporal punishment. Of thirty-four employees interviewed, twenty-six said they approved of corporal punishment, saw it as sometimes necessary, or thought it was not used frequently enough. Eight employees expressed disapproval, ambivalence, or no opinion about the use of corporal punishment on Samarcand inmates. Mary Lee Hunt, an elementary school teacher at Samarand, provided a representative answer to investigators when she "heartily" approved of corporal punishment in special cases of "last resort." She found that former teacher Arline James and nurse Viola Sistare "chummed with girls" and were of "harmful influence." Sistare, she complained, had "encouraged girls to talk imprudently to teacher [*sic*]." Investigators could not find girls willing to relate horror stories about brutal beatings. Two girls reported to investigators that "Altho switchings or whippings 'hurt,' they had not hurt too much." They thought the whippings "did them good" and "saved me from the penitentiary." It seemed that perhaps James and Sistare could not be trusted.[6]

Despite the conflicting ethics among staff regarding corporal punishment, everyone agreed on one general fact. Whipping, for good or ill, was a common form of discipline at the reformatory. Most of the teachers who told investigators that they approved of whippings still drew their income from Samarcand. Many staff members remained loyal to MacNaughton. They dismissed the claims of former nurses and teachers, such as Sistare, James, and Piland, who, they argued, may have sought retribution against MacNaughton by depicting her as cruel. Whose story should the historian favor, that of dismissed teachers or that of loyal staff? It is possible to gain some insight about discipline at Samarcand from interviews of other staff members when their comments are interpreted with an eye toward their biases. Some staff members echoed MacNaughton's complaints that the dismissed teachers were "harmful influences" on the girls. But others commented on the very good but misplaced intentions held by the dismissed staff.

Records indicate that the dismissed teachers and nurse had good intentions. Piland, James, Sistare, and Mitchem possessed a missionary zeal to save the girls physically and spiritually. Lottie Mitchem's "chummy" behavior, as one staff member noted, may have arisen from her desire to spread the gospel among her charges. William Crenshaw noted that Lottie Mitchem "singled out girls to make Christians of." Girls mocked Mitchem's biblical mottoes, such as "yield not to temptation" and "keep sweet." Teachers who wanted to spread Christ's message at Samarcand found no ally in Mrs. MacNaughton. Bessie Bishop, a nurse who left in 1930, complained of Mrs. MacNaughton's deep cynicism toward converting inmates to the church. She had never heard Mrs. MacNaughton offer any girl "a good heart to heart spiritual talk at all" and she would not allow the teachers to talk of it. "I was telling her one day of a girl I had talked to and she shrugged [*sic*] her shoulder at me and said, 'hah you had better be saving your breath' and I said 'but Miss MacNaughton God wants me to and that is my mission to try to turn them to Christ.' Again MacNaughton responded, 'ha you do not know them as I do.'"[7]

Most likely, regional conflict played a role in the creation of a tense atmosphere. At least two of the dismissed staff were from New England, including Viola Sistare. Maude Moore, a hall counselor, complained that Sistare had once announced, "I'm a Yankee—you Southerners get on my nerves." Physical plant manager J. H. Bodenheimer offered his own explanation for the teacher conflict. The four dismissed staff, he argued, were "from [the] North [and] were Catholics and ran together." While the women's faith remains unconfirmed and only one, Lottie Mitchem, exhibited any evangelical zeal, the comments by remaining staff indicate a clear regional clash between rural southerners and "Yankee" "Catholics." In fact, Bodenheimer may have used the term "Catholic" to apply broadly to any outsider.[8]

That some southern whites condoned the beatings of white children is consistent with the history of violent abuse in the South. Abusive and violent forms of punishment, especially as employed in the U.S. South, conjures images of the most pathological and violent character of slavery. Southern slaveholders and their legislators maintained this racialized order through extremely violent and oppressive practices to instill fear and submission in slave populations. However, the southern patriarchy also condoned and employed whipping and violent abuse to maintain the place of women of all classes. Throughout the nineteenth-century, coverture laws stipulated that a woman's body belonged to her husband. A woman could not bring charges against her husband for rape, and the courts forbade any woman from charging her husband with abuse and battery, as long as he used a stick "no larger than his thumb." Whippings, especially when used against women, were allowed. When abuse is placed in this larger historical context, it comes as no surprise that some southern employees at Samarcand were unfazed by corporal punishment at the institution. The practice of whipping white girls was consistent with their gendered and racial worldview.[9]

Teacher relations, corporal punishment, and treatment of venereal diseases ranked as the more high-profile issues in the investigation. Yet the issue with the greatest historical impact revolved around complaints of MacNaughton's parole system. Parole, the process that governed the decision to release a girl back to her county, rested with the Samarcand Board of Managers, MacNaughton, and her staff. County officials had no control over a girl's release. Mac-

Naughton's process stymied and frustrated court officials and county public welfare officers, who registered their own complaints to the investigation team. For years, county officials had complained in person that inmates were paroled from Samarcand and returned to counties without any notification. One county officer noted that one girl was returned to a home "not a fit place for a girl." Another official complained that he knew of a girl's return to her relatives only when she again landed in jail. Social workers complained that communication between the institution and county officials was nonexistent. Rumors abounded that girls from Samarcand were promiscuous and immoral. Tales that girls practiced "sex perversion" and "sodomy" and paired off with other girls as "sweethearts" preceded girls who returned home and bedeviled efforts of social workers who tried to place them in foster homes. One social worker complained that in her community a rumor had spread that the nurse in the Samarcand hospital had "produced" a miscarriage in one girl. Still further, none of the officials had a clear understanding whether paroled girls were free of venereal disease infection. Probably there was some truth to some of the rumors, yet the Samarcand authorities did little to clarify circumstances.[10]

Arbitrary parole policies worsened the communication between Samarcand and county officials. Statute permitted MacNaughton and the Samarcand Board of Managers full discretion in admission and discharge of inmates. Although magistrates sentenced girls to Samarcand, the administration could refuse admittance on a case-by-case basis, forcing county officials to resume care. The law permitted the staff to confine a girl at the board's discretion, but for no longer than three years. The investigation team discovered that generally in 1930–31 girls spent no longer than two and a half years at Samarcand. A small number spent more time at the institution. Five girls classified as "repeaters or girls breaking parole" had averaged "6 yrs, 4 mos, 24 days" in admission, parole, and readmission.[11]

MacNaughton's parole process rested on a system designed to reward girls for proper, ladylike behavior. A girl's parole depended on her accumulation of merit points earned for good behavior. For bad behavior, girls suffered demerits and an extended sentence. While staff also resorted to other disciplinary measures, including additional work, solitary confinement, and "spanking," parole depended on the merit system. Investigators did not specify the number of merit points a girl had to earn for parole. The exact number likely was fixed at the superintendent's discretion, but everyone knew the demerit rules and regulations. The "Rules and Regulations" chart was posted in the halls.

"RULES AND REGULATIONS"	DEMERITS
1. Leaving the hall without permission	3
2. Rudeness to anyone	2–5
3. Quarrelling—scrub basement	5
4. Appearance	3
5. Slang (2 call downs)	3
6. Loud talking (1 call down)	1
7. Discourteous to anyone	9
8. Visiting without permission (unless during visiting hours)	4
9. Talking after lining up and on sleeping porch	4
10. Uncleanliness (a visit to Miss Crenshaw)	8
11. Not getting up when called	6
12. Having to do household tasks over	(1 call down)

Any girl who does not lose any merits during one month will have a surprise at the end of each month.

The girls who lose ten merits or get ten demerits during any one month, means a visit to Miss Crenshaw (Supervisor Student Government)

13. Making cute unnecessary remarks	3
14. Misbehavior in Chapel	7[12]

The rules apparently rewarded submissive, quiet, and courteous girls. But the demerit system was unevenly distributed and redundant. As shown in the "Rules and Regulations" chart, the points system was unfair and open to interpretation. Girls who talked, depending upon circumstances (as outlined in rules 6, 9, 13, and 14), could lose anywhere between one and seven points toward parole. Girls who exhibited rudeness might lose two to five points (rule 2), nine points (rule 7), or three points (rule 13). And points docked for poor appearance (rule 4) or uncleanliness (rule 10) ranged from three to eight demerits.

Despite the rigid demerit system, some girls in fact earned parole. Unfortunately, county officials were not always aware of a girl's impending return home. The confusion not only proved inconvenient but also could land a girl in the same environment or worse than what had led to her confinement at Samarcand in the first place. Of 138 girls paroled in 1929, the investigation team found that twenty-four were sent home without notification to county officials, six arrived in another state without notification, and four appeared home before the notice arrived. The year 1930 proved only a little better. Of

173 girls paroled, twenty-five were sent home without notification and three appeared at home before notice arrived.[13]

The subject of parole mattered to the investigation team, but they buried the issue in their list of recommendations, a report that emphasized improved employee relations and use of solitary confinement, a style of discipline often used in modern adult prisons. In late May 1931, just following the trial, the State Board of Charities and Public Welfare presented a list of thirteen recommendations to the Board of Trustees for the State Home and Industrial School for Girls. Six of the recommendations involved salaries, staff hiring, record-keeping policies, and improvement of employee morale. One recommendation related to improving oversight by the Moore County school superintendent of education, and another addressed space in the infirmary. Two recommendations addressed punishment. Recommendation #7 banned whippings and encouraged the use of restricted diets, solitary confinement (later amended to the term "thinking rooms"), and written records of discipline on every girl. Recommendation #8 stipulated that more difficult or insubordinate inmates should remain at Samarcand "for longer rather than shorter periods," and that solitary confinement be used to "adjust" girls. A ninth recommendation discouraged admittance of girls considered "neglected, dependent, or homeless." Rather, Samarcand should house girls "who are delinquent, incorrigible, etc., and who cannot receive the treatment and training they need elsewhere in the state."[14]

Finally, two recommendations addressed parole. The investigation team demanded more accountability in the parole system and a thorough "post-release" plan. Recommendation #10 required a "classification or case committee," composed of the superintendent, assistant superintendent, and hall counselor, to review the credit and parole system, to record in written monthly supervisory reports each girl's progress, and to determine when a girl should be paroled. In addition, the board urged Samarcand to adopt a more methodical system of parole that provided a coordinated plan for each girl upon her release to county welfare officials. Paroled girls, the recommendations intoned, should not be left to the four winds.[15]

The recommendations prompted no radical reform, nor did the report criticize Agnes MacNaughton's leadership. Banning whippings proved the only immediate action of significance. Upon the report's release, the Board of Managers quickly passed a ban on corporal punishment. Otherwise, the investigation team praised the Samarcand officials for decades of good work. "In reviewing the work done at Samarcand," they concluded, "it is apparent that while certain changes in organization and policy are recommended, looking to improvement in these respects, the efforts of those in charge of administration of the institution since its establishment in 1917 have yielded rich returns in remaking and the reclamation of lives of many young women who have had the benefit of the training received there which after all is the only practical and final test to be applied."[16]

MacNaughton's implementation of parole, though inconsistent, reflected her belief that Samarcand's purpose was as a substitute parent that would set wayward white girls back on the path to southern ladyhood. Although aware that the girls might engage in serious delinquency and willing to punish by whipping, MacNaughton always held out hope that Samarcand served the state not as an asylum or prison, but as a training center in domesticity for the state's future white women. Generally, the investigation team agreed with her. On the question of corporal punishment they disagreed. Also, they suggested a more methodical and accountable system of classification, but did not specify the scope of such a plan. The team agreed that only a few reforms were necessary. These girls were trouble and required institutionalization, yes. But in 1931, no one—not the press, not the State Board of Charities and Public Welfare, not the public—called for the intelligence testing or sterilization of inmates.

The Board of Managers moved slowly on adopting new parole procedures. Inconsistent discharge policies continued to frustrate county social workers. Legal authority rested the fate of a girl's parole or release with Samarcand personnel. Sometimes, little communication passed between Samarcand and county authorities when an inmate was sent home. Social workers continued to complain of inadequate notice and lack of instruction regarding a girl's release. In the fall of 1932, Mrs. MacNaughton sent "Ethel" home to Forsyth County after three weeks because Samarcand "did not keep girls who wet the bed." The social worker sent the matter to MacNaughton's supervisor, Public Welfare Commissioner Annie Kizer Bost. MacNaughton's decision was arbitrary, she complained, and the girl unquestionably was an institutional case. Ethel, explained social worker Minnie R. Kimball, was a victim of sexual molestation. The court suspected that her grandfather and a "young boy" had used her for prostitution. She had tested negative for venereal disease, but Kimball referred to her anyway as a "sex case." "Most sex cases are bed wetters," Kimball concluded, and juvenile institutions were the only resource for treating such cases. Ethel had come from one of Forsyth County's "most intensive problem families," Kimball conceded, and "was definitely a case for them [Samarcand] to handle." The girl's father was incarcerated on a murder charge and the court had little choice but to commit Ethel to Samarcand. From the perspective of this county social worker, Samarcand should not have released the girl.[17]

The lack of a clear policy reflected badly on Agnes MacNaughton and the Samarcand Board of Managers. The director of the Division of Institutions, R. Eugene Brown, acknowledged to Kimball that unfortunately state statute vested in MacNaughton the discretionary authority of parole and release, thus no one could intervene. While MacNaughton's decision appeared "rather drastic," Brown explained to Kimball, "Miss MacNaughton may have had other reasons which she did not give for discharging this particular girl." Brown inquired about the case with MacNaughton, but only so far as to suggest that it should be customary to provide a community two weeks' notice before returning a girl. Kimball could hardly contain her annoyance. Two years previous, Samarcand had returned four girls at one time without any notification whatsoever. Kimball had carefully prepared MacNaughton for Ethel's admission with medical and psychological testing and a telephone conversation. In return, she expected Samarcand to notify her ahead of the girl's return home. But no notification came. MacNaughton's letter arrived twenty-four hours after the girl's release. The delay infuriated Kimball and prevented her the time she needed to help Ethel readjust to the community. When Ethel arrived in Winston-Salem, she immediately married the young man accused of pimping her. Social workers, Kimball reported, had no time to save the girl and to prevent the union.[18]

MacNaughton's leadership style, once praised as an example of progressive reform, was losing its luster in 1933. Her unapologetic defense of corporal punishment and her authoritarian approach to parole and discharge appeared inconsistent, lacking in process, and undemocratic. Quietly, the Board of Managers shifted to new leadership. The change in leadership was only intended as temporary, at first. When MacNaughton's health began to decline in late 1933, the Board of Managers appointed Grace M. Robson as acting superintendent of Samarcand Manor. It is unclear whether MacNaughton was asked to resign or did so of her own volition. Regardless, by July 1934 it had become clear that MacNaughton would not return. The board considered applications from social workers and corrections officers, but favored Robson, whose background reflected her training in nursing and mental hygiene. A graduate from the Woman's College Hospital in Philadelphia and the former superintendent of an institution for feebleminded women in Clinton, New Jersey, Robson represented the board's shift away from a policy that emphasized social reform or juvenile delinquency corrections. Instead, the appointment of a mental hygienist heralded for Samarcand inmates a new era of psychological testing, classification, and sterilization to a degree that previously had not existed at the institution.[19]

Grace M. Robson reflected the racial ideology of the mental hygienists. Her medical training and corrections background compared to the experiences of Kate Burr Johnson and Martha P. Falconer, women who previously had contributed to state eugenic policy. Robson embarked upon a plan to dramatically reshape Samarcand Manor. In doing so, she instituted the mental hygienist's racial policy by stripping poor white girls of their privileges of whiteness and "reclassifying" them as a separate, unfit population. First, she discharged the youngest Samarcand girls, those admitted under age ten as orphans. Second, she reformed the hospital so as to change its mission from a place of treatment for various ailments to a facility of psychological and venereal disease testing and treatment. Third, she instituted a complex classification program intended to separate the fit from the unfit and to determine a recommendation on sterilization. In accomplishing these steps she redesigned Samarcand Manor into an institution that would house and treat the white girls lost to the race.

Grace M. Robson inherited the responsibility for an institution with a progressively worsening rap sheet. The 1931 fire, the trial, and reports of rampant venereal disease among the inmates had contributed to a negative perception of Samarcand as an institution that housed girls who were irredeemable to society. Robson faced her first public relations crisis in April 1934 when the *News and Observer* published an article that cast Samarcand in its most negative light yet. A Raleigh *News and Observer* journalist, Herbert O'Keef, depicted Samarcand as a facility for social reprobates. O'Keef based his observations on little more than speculation and assumption taken from a visit to the institution. "Two hundred and fifty of the greatest problems in North Carolina are in Samarcand Manor down in Moore County," O'Keef began. These problems are 250 girls, every one of them admitted with venereal disease. The girls represented a major social threat, for if "allowed to remain in her home town, she will be a diseased prostitute and as such will be a material factor in the spread of venereal disease." O'Keef wagered that none of the girls had any chance at successful readmission into society. "Samarcand can never give the girl an equal chance," O'Keef argued, because "if it once becomes known that a girl has been at Samarcand, her hopes of living down her past are largely blasted."[20]

Social workers and welfare personnel greeted the article with indignation and outrage. L. L. Boyd, superintendent of Morrison Industrial Training School, North Carolina's state reformatory for African American boys, called for censoring or retracting the article. The information in the article represented "one of the rawest deals for helpless and defenseless inmates that I have ever read." Kate Capehart, a case worker in Union County, also demanded retraction on account of the misinformation on venereal disease infection. Capehart reported that only 30 percent of the girls in the institution were infected upon admission. She agreed that the population of girls at Samarcand represented a significant problem for the state, "but a false impression as to the seriousness should not be allowed to go unchallenged." The misleading information already had sparked a public backlash damaging to former inmates. One foster family had suddenly rejected a nine-year-old girl recently released from Samarcand. Capehart had successfully placed the girl in a "good home," but "as soon as that article was read by the family in question the little girl was discharged as a menace although she had never had the disease in any form."[21]

The article threatened to unravel the careful placement of twenty-six Samarcand girls under age ten (known as the "little girls") intended for placement in foster homes. The Board of Public Welfare had decided in its 1931 recommendations that as a juvenile delinquency institution, Samarcand Manor was an unsuitable facility for orphaned girls under age ten. Just one month previous to the publication of O'Keef's article, Samarcand had released twenty-six little girls under ten years of age. The institution had always kept a few inmates, some as young as six years old, who were admitted to the institution not as delinquents but as orphans. The nine-year-old "menace" in Union County most certainly was one of the discharged little girls. Grace Robson faced a public relations disaster if other families who had just taken in these girls also rejected them because of the charges in O'Keef's article. Samarcand and the Division of Child Welfare had spent months carefully coordinating the discharge, including six girls to families in Union County. O'Keef had observed these little girls and argued that they should remain at Samarcand. They were, he said, the one group "absolutely happy at being at Samarcand." Institutionalized due to parental neglect, they had come from homes where "prostitution was carried on, drunkenness commonplace." Not only were the girls the happiest of the lot, he claimed, but they should remain at Samarcand because "each of these 8–12 year old children had venereal disease when admitted." This inaccurate and inflammatory statement alarmed the foster family in Union County, who had been promised that the girl was healthy.[22]

In the 1920s the presence of the "little girls" in photos and descriptions of Samarcand Manor had contributed to an image of wholesomeness that lent an air of innocence to the institution. However, after the 1931 trial the Board of Public Welfare had agreed that for the sake of consistency in housing, education, discipline, and parole, the institution should serve only girls convicted as delinquents and not as a home for "neglected and dependent children." Under the direction of Lily E. Mitchell, the director of the Division of Child Welfare, Superintendent Robson attended to the subsequent discharge of the girls by securing medical and psychological evaluations of each and providing proper notification to the girls' home counties. Medical tests confirmed that some of the little girls were infected with venereal disease, which authorities attributed not to "sex delinquency on their own part, but through neglect of parents and improper environment." They were, like other girls, routinely treated and monitored. In an attempt to stave off the bad press, Robson submitted to her supervisor the statistics reflecting the actual incidence of venereal disease among girls admitted to Samarcand. Of 131 cases admitted from May 1, 1933, to May 1, 1934, she explained only thirty-three (25 percent) were positive. "I am giving you this information," she informed Bost, "in order that you may make use of it in any way you see fit in combatting this re-action, which we are receiving from the publicity." Interestingly, no one pursued the question of how girls under age ten had contracted venereal disease in the first place.[23]

Once Robson had fostered or placed out the "little girls," she began to streamline admittance procedures. The admittance record of "Eliza" typified a girl's orientation at Samarcand. A girl's first stop was the hospital. Eliza's social worker had accompanied her all the way from Onslow County to Eagle Springs, nearly two hundred miles, upon her commitment by the Onslow juvenile court for "dependency." They arrived first at the administrative building. Grace M. Robson welcomed them and advised her that she would spend the next ten days in quarantine and isolation at the hospital, where she would undergo several medical and mental evaluations, routine procedure for every new girl admitted at Samarcand. The social worker departed for home and left Eliza in the hands of the nurse, the only full-time medical staffer at the hospital. Accompanied by the nurse, Eliza exited the rear of the administrative building and walked to the two-story hospital, only a few paces from the main building. Upon entry, she noted the large living room and fireplace on the first floor. A sewing class, consisting of several other hospital patients, was under way on the glassed-in porch. Eliza was admitted to a second-floor quarantine room, consisting only of two beds, one of which was assigned to another new girl. Later she would find the basement, described by one observer as presenting a "somewhat dreary appearance," which served as the dining room and the kitchen. One can imagine Eliza's anxiety and anticipation as she readied herself for ten days in this place. This visit would begin her two years at Samarcand Manor.[24]

The hospital at Samarcand had functioned since the school opened in 1917, and it served girls with all sorts of afflictions. A significant part of its operations always had been quarantine, isolation, and treatment for venereal disease. In World War I, the federal government, intending to protect U.S. troops stationed in North Carolina, had provided significant funding to treat girls with syphilis, viewed as a threat to the public health. Physicians also used the facilities to treat common ailments and to perform minor surgical procedures, such as tonsillectomies and appendectomies. Nurses performed psychological evaluations and treated girls bruised by disciplinary whippings. In 1934, the physician in charge, Dr. J. P. Bowen, gave medical instruction for the care of venereal disease at Samarcand. He instructed staff to quarantine all patients immediately upon diagnosis of any "luetic, gonoccal [sic] or chaneroidal [sic] infection" and exhibiting "primary or secondary lesions" until pronounced noninfectious by the attending physician. Any girl showing a positive Wasserman reaction (for syphilis), regardless of the presence or absence of lesions, would be isolated and treated as well. Once diagnosed with disease, no patient would be allowed "to prepare, serve, transport, or handle in any way foodstuffs, bed linen, clothing, medical supplies, bathroom supplies, kitchen utensils, or dining room supplies to be used by others not infected with her particular disease until she has been pronounced non-infectious by the attending physician."[25]

By 1934, under Robson's authority, the hospital took on a more nefarious purpose. Not only did it serve as a quarantine facility and cottage for girls afflicted with venereal disease, but also it was the first link in the chain of decision-making that led to a girl's sterilization. These fateful decisions progressed through the new system for classification and parole evaluations. Staff administered psychological evaluations that would determine a girl's treatment, residential placement, form of education, and training. This data, compiled with other behavioral reports and evaluations, would serve as the heart of the material that staff would use for decisions about sterilization and parole. A state psychologist visited Samarcand to administer intelligence quotient tests, including the Stanford Binet, a Porteus Maze test, and Healey's Boys Day test. In the spring of 1934, only twenty-nine of 170 girls tested met an IQ score of eighty or above. A girl's score was significant. Not only did it determine her level of training and her residence hall assignment, but it also determined her reproductive fate. Institutional policy recommended sterilization of all girls who tested below an IQ of sixty-five. In the spring of 1934, of sixty-two cases recommended for sterilization, the state had performed thirteen operations. Six more were pending approval and eight were awaiting hearings before the state eugenics commission.[26]

This change in direction for the hospital grew out of Robson's leadership as the superintendent who replaced Agnes MacNaughton. The board appointed Robson in 1934, but she had begun politicking her own approach to institutional treatment in 1933, while she served as interim superintendent. In a 1933 address at the Fifteenth Annual Public Welfare Institute in Chapel Hill, North Carolina, Robson argued that institutional care at juvenile reformatories required a modern and methodological system of classification and parole based upon practices in social hygiene. "The newer purpose of the Institution," she argued, "makes it imperative that the Institution utilize all that is known to us of scientific methods of dealing with causal factors and the treatment of individual delinquents." The new process would require a child's observation, evaluation, and reporting by numerous experts in medicine, psychology, and social work and by institutional staff. "It is no simple task, this one, of untangling a complex of emotional and physical symptoms," observed one university professor of social work. "The insight into each girl's problems—mental, physical and social—enables the guidance personnel to help where she needs help most, both individually and in her life with the group."[27]

The scientific method provided a routine plan of diagnosis, treatment, and parole that not only determined a child's treatment within the institution, but also dictated the conditions of parole and a child's release back into the community. Robson rejected traditional progressive reform policies for the down-and-out class. She articulated her plan in the language of mental and social hygiene. Based upon methods in science, the three-phase process mimicked a clinical approach including diagnosis, treatment, and aftercare. The first step, "Diagnosis," was defined as "the classification [by experts] of each child for the purpose of diagnosis, and setting of goals to be attained if possible." The second stage, "Treatment," involved "assisting each child to attain the individual goals, and preparation of the environment for return of the child to the community when goals are attained." The third phase "After Care or Parole," also referred to as "readjustment," necessitated the "determination of the child's fitness for returning to family life in the Community." The ultimate goal was a holistic one intended to shape a child's future adult life. The child, Robson stressed, "is to be taught to become a useful citizen and not to be just a good child in a daily routine of Institutional life."[28]

The decision to sterilize an inmate rested at the heart of this process. In practice, the classification committee, a team of experts designed to enact the three-phase program, met periodically to evaluate each girl and discuss not only her care and training, but also her reproductive fate. Robson considered the issue of sterilization and its implementation as an important component of this classification system. In fact, it might be said that the committee designed the classification system almost solely to distinguish between those girls who were feebleminded and sexually "active" and those who were not. Medical examinations chiefly reported on whether or not a girl had venereal disease, hardly mentioning other medical issues. Psychological examinations only tested her intelligence quotient and commented only on her degree of "feeblemindedness." According to the *Public Welfare News*, "many [inmates] are below the mental level at which social adjustment in a complex society will ever be possible . . . a small percentage are of the imbecile level and cannot fit into a school plan, except in a custodial long-time training program." Other observations reported on her academic or vocational potential and her general demeanor with staff and other inmates. In other words, in reviewing each girl's file, the committee focused chiefly upon the questions: Is she feebleminded or not? Is she sexually active or not? Should she be sterilized or not?[29]

The order to sterilize rested upon a girl's psychological evaluation. Samarcand employed several intelligence tests. The most recognized was the Binet test. Henry Goddard introduced the Binet intelligence test in 1908. The subjective nature of the test stoked controversy among both experts and laypeople. Nonetheless, experts upheld the Binet as scientifically valid and reliable in measuring an individual's mental intelligence and their psychological strengths and weaknesses. Psychologists employed the test at the Illinois State Training School for Girls at Geneva, an institution very much like Samarcand. Both reformatories housed "feebleminded" girls. The difference in the two institutions is that Illinois did not permit sterilizations. Michael A. Rembis has examined individual psychological evaluations and scores based upon the Binet test performed after World War I. The test process was always fluid and dynamic (and therefore very unscientific). Testers often linked mental acuity to subjective interpretations of a girl's actions and behavior in the course of an interview. Testers applied their preconceived notions of "normal" intelligence in their concluding reports. Their reports also might judge a girl who performed well upon other unrelated behavior, such as her abilities as a "good dancer." Psychologists who performed the exams generally perceived lack of intelligence as linked to inappropriate social behaviors. Girls who performed poorly might exhibit "sex experience" or seem "easily flattered." Testers linked mental defectiveness to quarrelsome or garrulous behavior, anxiety, pregnancy, and venereal disease. Sometimes evaluators noted a girl's test anxiety, physical condition such as pregnancy or venereal disease, and history of sexual abuse, but, they claimed, these indications did not influence any outcome in the tests. At the Geneva school, testers found 95 percent of girls to be feebleminded. The testers concluded that "low grade intelligence" was the source of immoral behavior in 58 percent of the inmates.[30]

Robson probably employed the Binet test in a similar fashion. Testing and sterilization were not forms of punishment. Robson viewed testing as a clinical step toward sterilization, a social necessity, and a precondition of discharge or parole. Feeblemindedness represented an undesirable genetic trait sometimes linked to criminality. Robson argued that no single factor, such as "feeblemindedness, poverty, bad sex habits," or "gang spirit," produced a delinquent or criminal. "It is rather," she remarked, "the whole background of experience plus his mental capacity and constitutional makeup that determines his attitudes in life and his reactions to situations." Institutions could correct many bad behaviors, but feeblemindedness was the one characteristic that the institution could never address through readjustment, rehabilitation, reeducation, or reform. The classification committee, she said, thus should recommend sterilization prior to parole for all offenders "likely to propagate mentally unfit offspring." Also prior to release, the social worker to be responsible for parole supervision would present a report on community resources for pre-parole review. Transition back into the community was fraught with hazards, and required careful supervision by social workers. [31]

State-sponsored sterilization programs originally targeted young white females. [32] In the South, concerns about white supremacy and class conflict drove the movement to control the reproductive and sexual freedoms of poor men and women, both black and white. Sterilization policies targeted young working-class women because authorities found this group easiest to control. Forced sterilization, in the guise of science and protection, was a form of state-sponsored violence and deception used toward enforcing patriarchal and racial dominance over working-class girls. Virginia was the first southern state to adopt the practice, in 1924, when officials sterilized Carrie Buck, a Charlottesville teenager whose adopted family committed her for feeblemindedness after finding her pregnant, a result of rape. In 1927 the U.S. Supreme Court sanctioned Virginia's law permitting sterilization of epileptic, feebleminded, and insane persons judged a danger to themselves or others. Experts had ruled that Annie Buck, her mother, and Carrie belonged to a "shiftless" class of southern whites. Oliver Wendell Holmes ruled that "three generations of imbeciles are enough." By the mid-1930s, thirty states had adopted sterilization laws as a result of the *Buck v. Bell* ruling. [33]

At first it was not so easy for a public welfare official to demand sterilization of a girl. The state's first sterilization laws required a political process that deterred public welfare authorities from making the case. North Carolina's first sterilization statute dates to 1919. The law did not make use of the word sterilization, but allowed penal and charitable institutions to petition for involuntary sterilization where it might "benefit the moral, mental or physical conditions" of inmates. Some public welfare authorities referred to it as "mechanically cumbersome," probably because it required that the governor sign each sterilization order. Public Welfare Commissioner Kate Burr Johnson pressured legislators to pass a new law in 1929 that would empower public welfare officials to pursue forcible sterilizations in their counties. This law specified sterilization for the benefit of the health of the county poor, but also added that it should be allowed in instances "for the public good." The law called for sterilization of "mentally defective" patients when approved by the state health officer, the superintendent of public welfare, and two state physicians or by the next of kin. State senator H. L. Milner of Burke County sponsored the law.

He reminded legislators of the need for "pure bred stock," "pure bred seed," and "pure bred people." His bill, he argued, would "stop the propagation of children by those who are suffering from defective mentalities of the sort that would be inherited by their children." Senator Galloway objected, saying that the practice was "proven false by Asiatics in the days of eunuchs." But doctors on the floor argued the practice was safe and preserved an individuals' ability to feel or satisfy sexual desires. The 1929 law did not survive scrutiny in the state supreme court. In 1932, the North Carolina Supreme Court held the law as unconstitutional because it provided no notice of hearing or right of appeal.[34]

The General Assembly tried yet a third time to legalize forced sterilization in 1933. Representative W. A. Thompson of Beaufort County and a member of the board of directors for Caswell Training School, an institution for the feebleminded, proposed the new law so as to avoid further legal challenges. To provide for hearings and appeals, the 1933 law created the Eugenics Board to oversee sterilizations. This board consisted of the commissioner of public welfare, the secretary of the Board of Health, two state physicians, and the attorney general of North Carolina. The board accepted petitions from the state institutions and from county public welfare officials. The North Carolina process represented the only one in the nation that gave social workers power to file petitions. The new sterilization law provided for sterilization of any mentally diseased, feebleminded, or epileptic inmate or patient of the state or county institutions or any other residents of any county where the county commissioners believed sterilization was in the best interest of the public good. It also allowed sterilization when social workers thought an individual or inmate would procreate a child with a tendency toward serious physical or mental deficiency. The *News and Observer* heralded the law as in the interest of humanity. "More and more in our times large families are disappearing in the households of the men who do the world's work and pay the world's taxes." Fearing a decline in this so-called desirable "human stock," the *News and Observer* called for sterilization of the "lowest orders of humanity" in the population: "We cannot make a better world if we deliberately give our substance to subsidizing the production of the least worthy stock among men."[35]

In summary, Robson had the support of the legislature for sterilizations by 1933. Begun in 1919, the process of sterilizing inmates was a political one and required the signature of the governor. Few sterilization orders passed. A 1929 law streamlined the process, but was declared unconstitutional in 1932 due to the lack of an appeals process. In 1933 the legislature amended the law, removed the governor from the process, and rested the decision-making on the institution, the State Board of Public Welfare, and a newly formed Eugenics Board. Further modifications to the law in 1935 placed even more power into institutions' classification committees, which began to send "many applications for eugenical sterilization to the Eugenics Board." According to reports by the State Board of Public Welfare, "most of the cases recommended are both mentally and socially feebleminded, and in most instances, the Eugenics Board accepts the recommendation for sterilization operation."[36]

The 1933 law conspicuously cited the German sterilization law of the Nazi regime. And it resembled it too. Both laws subordinated the interests of the individual to that of the community by applying the law not only to those in institutions but also to the population at large, and sanctioned the use of force. Stefan Kuhl has shown that before 1933, eugenic laws in the United States influenced German eugenicists. Germany's 1933 "Law on Preventing Hereditarily Ill Progeny" was based on the California sterilization law. In turn, American eugenicists, such as some in North Carolina, admired the Nazi racial hygiene law of 1933. The German law differed in that it applied to individuals with other physical handicaps besides epilepsy, and with other undesirable traits, including drug and alcohol addiction and "sex delinquency." North Carolina's law generally applied in cases of epilepsy and sex delinquency. Both the German and North Carolina authorities attributed "sex delinquency"—that is, evidence of immoral or socially unacceptable behavior, promiscuity, or venereal disease in girls and women—as evidence of weak-mindedness, "low grade intelligence," and inability to control the "sex instinct."[37]

After World War II, North Carolina eugenicists made only halfhearted attempts to distance the North Carolina sterilization law from the Nazi regime. In 1950, sociologist Moya Woodside made a significant effort to distance the North Carolina statute from the Nazi legislation. The Nazis, she argued, applied the law far more broadly and provided a very short and restrictive appeals process, thus employing eugenics as a "potential weapon of class discrimination." The North Carolina statute bore no such resemblance, as it made provisions for safeguarding individual liberty. She did acknowledge, however, that in actual practice "it must be added that theoretical concepts of individual liberty are often remote from the sort of practical situation in which sterilization is usually proposed." In other words, she conceded that few officials considered the inherent rights and protected liberties of the feebleminded and sexually delinquent when they proposed sterilization.[38]

How did Samarcand staff select sterilization victims? The classification committee made the recommendation during a girl's "pre-parole review." The committee compiled a one- or two-page summary of tests and observations, most of which focused on the physical and mental capacity of the girl in question. The January 15, 1935, pre-parole review of "Ida" listed her birth, county of origin, date of admission, and cause of commitment, in this case defined as "dependency." Results of her social, medical, psychological, and other observational evaluations followed. By all accounts, Ida was healthy. Fifteen-year-old Ida was physically fit, had shown "marked improvement," had no "sex charges against her," was trained in "home making," and had no signs of venereal disease. However, the committee concluded she was feebleminded and that her father drank "excessively." Staff members recorded Ida's behavioral reports, best described as "lukewarm." Apparently she was "rather quarrelsome," but not too difficult to control, as she "is the type who follows rather than the one who instigates trouble." She was clever, and could work with her hands, especially when sufficiently supervised. The committee listed her "mental age" at ten years old and her Intelligence Quotient score at sixty-seven. Diagnosed as feebleminded, she was recommended for vocational training under close supervision. The committee showed apathy and a callous indifference to the question of sterilization, as Ida's score, at sixty-seven, marked her as a borderline case. She "could be sterilized or not," the committee concluded, as if it did not desire the responsibility of recommending the operation. With a degree of recklessness, they left the decision in the hands of the members of the Eugenics Board, who did not know the girl. Ida would do well, the committee concluded, if upon release her social worker carefully supervised her in a housekeeping job where she might earn a small salary.[39]

The insensitive and unfeeling disregard that the classification committee expressed of Ida's fate casts a dark shadow on the broader pattern of sterilization at Samarcand Manor and in North Carolina as a state. The data demonstrates that in North Carolina, state-sponsored sterilization operations targeted white females age ten to nineteen in the years before 1950. Historians have shown that most procedures were performed on inmates without full and informed consent of the inmate or her guardian. Table 1, "Sterilizations by Gender and Race, 1929–1950," shows that the Eugenics Board sterilized from 239 to 468 people per biennial term. The vast majority, 53–76 percent, were white females. A very significant percentage, approximately 40 percent, were female juveniles between the ages of ten and nineteen. As shown in table 2, "North Carolina State Sterilizations (Ages 10–19), 1929–1950," between 1929 and 1936 the Eugenics Board performed sterilization procedures on 332 people, 135 (41 percent) were females aged ten to nineteen years old. Adolescent females accounted for 29 percent to 47 percent of all sterilizations per biennial term. In biennial reports beginning in 1936, the state reported sterilizing 239–468 individuals per term. Nearly half of all state sterilizations were performed on girls aged ten to nineteen. Table 3, "Sterilizations of Samarcand Inmates, 1929–1950," demonstrates that 12 percent (293 of 2538) were performed at Samarcand Manor or on girls from Samarcand Manor.[40]

Medical and public welfare authorities keenly understood the potential controversy that might rest in the public mind about nonconsensual sterilization of young white girls. They justified the operations by characterizing victims as "subnormal." In 1950, Moya Woodside, a social sciences research assistant at the University of North Carolina at Chapel Hill, published a study entitled *Sterilization in North Carolina: A Sociological and Psychological Study.*

TABLE 1. Sterilizations by Gender and Race, 1929–1950

YEAR	MALE (W)	FEMALE (W)	MALE (N)	FEMALE (N)	TOTAL STERILIZATIONS	% FEMALE (W)
1929–36	24	253	37 (21C)	18	332	76%
1936–38	19	204	39	34	296	69%
1938–40	45	186	38	35	304	61%
1940–42	43	179	41	76	339	53%
1942–44	28	151	33	57	269	56%
1944–46	37	169	4	29	239	71%
1946–48	50	188	1	52	291	65%
1948–50	65	291	5	100	468[a]	63%

Source: Data from "Biennial Reports to Newspaper Clippings," Eugenics Commission, General File, 1933–1974, box 4.1. Data compiled from Appendices, 1934–1950. North Carolina State Archives. Raleigh, N.C. See also *Biennial Report of the Eugenics Board of North Carolina*, North Carolina History of Health Digital Collection, UNC-Health Sciences Library.

Note: [a] 1948–50 total includes sterilization of seven Indian females.

TABLE 2. North Carolina State Sterilizations (Ages 10–19), 1929–1950

YEAR	MALE (AGES 10–19)	FEMALE (AGES 10–19)	TOTAL (AGES 10–19)	ALL STERILIZATIONS	% FEMALE (AGES 10–19)	% ALL (AGES 10–19)
1929–36	14	135	149	332	41%	45%
1936–38	24	138	162	296	47%	55%
1938–40	21	130	151	304	43%	50%
1940–42	41	124	165	339	37%	49%
1942–44	32	117	149	269	43%	55%
1944–46	24	97	121	239	41%	51%
1946–48	31	101	132	291	35%	45%
1948–50	29	136	165	468	29%	35%

Source: Data from "Biennial Reports to Newspaper Clippings," Eugenics Commission, General File, 1933–1974, box 4.1. Data compiled from Appendices, 1934–1950. North Carolina State Archives. Raleigh, N.C. See also *Biennial Report of the Eugenics Board of North Carolina, North Carolina History of Health Digital Collection,* UNC-Health Sciences Library.

TABLE 3. Sterilizations of Samarcand Inmates, 1929–1950

YEAR	SALPINGECTOMY	OVARIECTOMY	SAMARCAND TOTAL	ALL STERILIZATIONS	PERCENT AT SAMARCAND
1929–36	5	4	9	332	3%
1936–38	42		42	296	14%
1938–40	64		64	304	21%
1940–42	36		36	339	11%
1942–44	57		57	269	21%
1944–46	40		40	239	17%
1946–48	24		24	291	8%
1948–50	21		21	468	5%
TOTAL:					
1929–50			293	2538	12%

Source: Data from "Biennial Reports to Newspaper Clippings," Eugenics Commission, General File, 1933–1974, box 4.1. Data compiled from Appendices, 1934–1950. North Carolina State Archives. Raleigh, N.C. See also *Biennial Report of the Eugenics Board of North Carolina*, North Carolina History of Health Digital Collection, UNC-Health Sciences Library.

Sterilization, she argued, was an effective means by which the state should control the "subnormal" population. Despite the public stigma associated with the process forged in part by its association with German social engineering during the Nazi era, she argued that state-sponsored sterilization nonetheless marked a progressive trajectory in social engineering. Woodside contemplated the very negative association of the term "subnormal" when applied to sterilization. She mused that the state should mitigate the public stigma that "subnormal" would attach to the procedure. The state, she proposed, should defeat the public stigma by also encouraging use of it in the medical community for reproductive relief in normal, married women. Such a course of action, she believed, would thus develop a positive association in the public mind.[41]

Yet Woodside shelved many of her reservations upon observing Samarcand girls. She described Samarcand inmates as "subnormal" and prime candidates for sterilization. In 1947 she visited the school and found two hundred girls, 50 percent of whom she identified as committed for sex offenses, but "mainly those associated with vagrancy and running around with boys." Others were committed for stealing, truancy or "minor delinquency indicative of lack of parental care." Of prostitutes, she commented, there were very few at the school. Many girls were orphans or had parents who were "drunken," feebleminded, and on public relief. The intelligence level of the girls averaged from sixty to eighty on the I.Q. scale. Of the more intelligent, Woodside observed, many tended "to be extremely unstable and unlikely ever to make a good adjustment in the community." A considerable number of the girls, she concluded, were "subnormal."[42]

"Subnormal" is not only a heartless term to describe these girls, but also inaccurate. It is true that many of the girls had little schooling. Most were probably naive, physically and sexually abused, and, as a result, suffered from serious psychological problems and low self-esteem. The mid-twentieth century South had little patience or interest in reclaiming a troubled youth, particularly a female. She would already have been seen as "lost." Such was the case with Melba, an eleven-year-old girl from a financially secure family in Mecklenburg County. Melba entered Samarcand by will of her parents. She had been a problem to Mecklenburg officials for some time. She was underweight, poorly developed, and was "extremely nervous." In other words, she suffered from anxiety. She had a history (described by Samarcand staff as a "mania") of stealing purses, often visiting schools and churches for the purpose of stealing. For these reasons, and by the suggestion of a county psychologist, her parents committed her to Samarcand.[43]

Melba's behavior at Samarcand suggests that she was badly abused, both sexually and physically, by someone at home. She entered fifth grade but did not adjust. Her behavior was erratic, and she misbehaved daily. According to the report, she suffered from extreme anxiety. Staff reported facial twitchings, bitten fingernails, and unsteady hands. She "was totally unable to keep her mind on her work long enough to grasp anything." She quarreled with other girls, left her seat without permission, and disobeyed commands and ignored requests. She demonstrated aggressive behavior in the classroom such as kicking desks and kicking over globes.[44]

Samarcand staff decided they could do little with her and so sent her home for a "vacation," releasing her in May 1934 for the summer. Robson chose to send her home because Melba was committed by her parents, not the juvenile court. When she returned, the staff noticed that Melba was in a "more run-down condition" than before. Thinner, dirty, neglected, they nonetheless allowed her to return home at Christmas for a few days. Finally, Grace Robson seemed to get the point. She was being abused at home. Melba was treated for gonorrhea. Diagnosed with an eye condition, she was prescribed glasses and placed on a restricted diet to treat her "nervousness." Her parents had asked to send her to a boarding school, but Robson disallowed this idea, prohibited any more home visits, and removed her from the academic setting into the vocational school. It is unclear what became of Melba, as records of individual girls remain closed. When she reached adulthood, she most likely was institutionalized at one of North Carolina's hospitals or sanitoriums for insane women. It is unknown whether or not she was sterilized. What happened to her after Samarcand remains unclear. In her tragedy we learn important historical lessons about how authorities viewed troubled white girls. Any modern reading of her file would likely conclude that she had post-traumatic stress disorder. A classification as "subnormal" appears ludicrous by today's standards.[45]

Yet in 1935 public welfare officials such as Moya Woodside classified Melba and girls with similar histories as "subnormal." Woodside argued that the defectiveness ran in families. The family of Ella Mae W., sent to Samarcand at age thirteen for truancy, stealing, and running the streets, offered a typical profile of a "problem family." Diagnosed as borderline mental deficiency, she was reported as "unstable, uncooperative, sullen and untruthful." Rather than seek environmental causes of her behavior in her personal experiences, Woodside attributed her behavior to her genetics. Her father was an itinerant peddler. Her mother, Woodside reported, was "highly nervous," quarrelsome, "beats the children" or "indulges them," is a poor housekeeper, does not get along with her husband, and "drinks to excess." Her eldest brother already was committed to an industrial school. Woodside observed that many of these traits recurred in "subnormal" families. Traits of institutionalization, prostitution, venereal disease, physical abnormalities, or drunkenness often went unnoticed in adults who were not criminal or a public nuisance. Only through their "misbegotten" children, Woodside concluded, did public officials discover the defectiveness.[46]

How did the state proceed to sterilize young girls from these families? According to Woodside, the process began with a girl's first classification conference. Staff compiled her medical, psychological, and school reports with behavioral reviews, IQ tests, and the data collected about her social history. The staff might recommend any girl for sterilization, but screened girls for the procedure from among those who tested below seventy on the I.Q. test. Superintendent Robson next passed the recommendation to the superintendent of public welfare in the county where the girl resided. Law required the county to specify a doctor, seek parental consent, and forward the case to the Eugenics Board. Woodside reports that parents were often apathetic, though some requested meetings with the superintendent to discuss the reasons for sterilization. Where parents would not sign, the state had the power to enforce by compulsion in a special hearing of the Eugenics Board. The superintendent explained the process to the girl in question, to avoid the likelihood of her parents "dissuading" her. "Most take it well and are not emotionally upset," Woodside commented, "but a few are afraid of the operation itself."[47]

The operation might be a bilateral tubal ligation (tubes tied), a bilateral salpingectomy (removal of both fallopian tubes), or an ovariectomy (removal of one or both ovaries). Girls were sent to a hospital in the county of their residence, usually upon parole. All costs rested with the county of residence. Some girls were sterilized before their release, at the local hospital in Pinehurst. Sterilization in Pinehurst, authorities argued, allowed convalescence at Samarcand Manor, reduced public exposure of the procedure in the girl's home county, and demonstrated to other girls at Samarcand that there was nothing "dreadful" about the surgery. A normal recovery required ten to fourteen days. Medical data is protected from researchers by state legislation, thus no specifics about medical procedures upon patients (adult or juvenile) are available. However, aggregated data from the biannual reports of the Eugenics Board demonstrate that nearly all of the sterilizations involved removal of the fallopian tubes. The Eugenics Board reported only four ovariectomies among the 293 operations performed on girls at Samarcand.[48]

White adolescent females at Samarcand thus endured a particularly disturbing form of sexual scrutiny before 1950. The classification committee routinely recommended girls for sterilization. In four consecutive meetings held in late 1934 (October 19, November 23, December 7, and December 21) the classification committee conducted pre-parole reviews of fifty-three girls. Each report briefly summarized the test results, goals, and recommendations discussed for each inmate. For example, on December 21, 1934, the committee reported an intelligence quotient of sixty-five for Clara Gurganus and recommended sterilization—"but not until next summer," her intended date of parole. The committee decided to leave another inmate, Cleo Tate, in academic classes because she would "benefit much by Samarcand training." The committee anticipated another pre-parole review for her in eighteen months, and recommended sterilization just before her release. Some of the sterilization recommendations appear as disciplinary measures. Rosalee Edmisten would not parole until May 1935, and should not be sterilized unless she violated her parole. Of the fifty-three girls reviewed in the four meetings, the committee recommended to sterilize nine (17 percent).[49]

Before 1933, no one seriously considered sterilizing girls at Samarcand. Agnes MacNaughton, despite her unapologetic defense of corporal punishment, never pushed for sterilization. The aging social reformer always held out some hope of rehabilitation of her charges. The move toward sterilization of girls, both inmates and parolees, occurred under the leadership of Grace M. Robson, a trained nurse and mental hygienist. The 1931 arson case was not the immediate cause or reason for the shift. Rather, the trial and the resulting investigation served as a catalyst toward greater scrutiny of the parole system. The trial contributed to Samarcand's poor public reputation, for certain. Yet, in hiring Grace Robson, the Board of Managers seemed to have made the conscious decision to move from a reformist approach to a mental hygiene agenda.

By the 1950s, eugenic science declined nationwide, but the policy goals remained. Although waning elsewhere, the procedure gained popularity in North Carolina under the leadership of Ellen Winston, who served as state commissioner of public welfare from 1944 to 1963. Winston, a student of the cultural sociologist William F. Ogburn, began a shift away from institutional sterilizations and aggressively promoted noninstitutional cases that targeted Aid to Dependent Children (ADC) recipients, who conservatives and liberals alike blamed for draining public funds. The public associated ADC with African American single mothers, who nationwide accounted for 48 percent of welfare recipients in 1961. Public discourse about ADC blamed black single mothers for illegitimacy, poverty, and social unrest. As early as 1953, Ellen Winston instituted a policy change in decisions of sterilization made by her office. "Special emphasis," she noted at the top of a list of new priorities toward ADC recipients, "is being placed upon sterilization of aid to dependent children mothers who come within the law." In 1957 and 1958, African American women constituted the majority of those sterilized by the state. North Carolina sterilized over eight thousand people between 1929 and 1975. In that period, the majority of victims were white women, but after 1958 most were African American. Winston was well rewarded for her service. In 1963, U.S. president John F. Kennedy appointed her the first U.S. commissioner of welfare in the newly created U.S. Department of Health, Education, and Welfare. In 1973, two women sued the state of North Carolina, a process that gained momentum and, in 2002, led to a high-profile report in the *Winston-Salem Journal*, an apology in 2003 by Governor Mike Easley, the creation of an "Office of Justice for Sterilization Victims" in 2010, and, in 2013, a claims process for compensation for victims.[50]

Grace M. Robson served at Samarcand from 1934 to 1944. With her at the helm, the institution took on new purpose. Suddenly, a chief objective of the reformatory was to identify and control the reproductive lives of the "subnormal." Robson had instituted a new classification and parole system designed to categorize girls by intelligence. The new program replaced the inconsistent merit or credit system that girls once followed to earn parole. Samarcand authorities used the new data to identify feebleminded or "subnormal" girls and recommend their sterilization to the Eugenics Board as a condition of parole. The new parole system marked a decisive shift in the institutionalization and treatment of juvenile white girls. The mental hygienist agenda of purifying the race had crushed the reformers' progressive impulse of rehabilitating bad white girls into good ones. Both agendas were steeped in white supremacist ideals, but one was a considerably more modernized racism than the other. This modern form of racism was rooted in a belief that whiteness consisted of superior characteristics unlikely to survive and propagate in the white poor. The state treated this class as another race, and targeted the girls and women of this class as subjects of possible sterilization. From 1933 to 1947, the State Eugenics Board sterilized 293 girls at Samarcand who failed to measure up to the standards of white purity.

8

"A Mystery to Me"

THE PROBLEM OF INCARCERATING FEMALE DELINQUENTS
IN WORLD WAR II-ERA NORTH CAROLINA

TWELVE YEARS HAD PASSED SINCE the Samarcand arson case. Judge Frank M. Armstrong sat on the bench of the Moore County Court, the courthouse that had tried the arson defendants. He had not committed a girl to Samarcand Manor in months. By 1943 the institution had gained such a bad reputation that reformers and advocates of even the most incorrigible girls had pleaded for sentences to workhouses and the state penitentiary instead. Armstrong refused to send girls there because he had heard rumors "that all sorts and forms of sexual perversion exist at this place." One terrible tragedy, a juvenile case, had presented him with a significant challenge. The state had charged a Winston-Salem girl of infanticide. The prosecutor had argued that upon the baby's birth, the defendant had hit her infant's head against the wall and killed it. She proclaimed innocence, saying that upon delivering the baby alone and in secret, the child fell to the floor and crushed its skull. The court found her guilty for "destroying her baby at birth." Upon her conviction, the clubwomen of Winston-Salem rallied to her defense and claimed she was a "nice girl" who should not suffer her sins at Samarcand. A prison sentence, they pleaded, was the more merciful alternative. Judge Armstrong concurred and sent her to prison. He had sentenced a child as an adult.[1]

Why did Armstrong, a juvenile court judge authorized by the juvenile courts provision of the Child Welfare Act of 1919, commit a girl as an adult? This chapter examines the public discourse on white womanhood and sexuality that surrounded women and girl inmates in the 1940s. As courts invoked juvenile reform to supervise, monitor, and correct the behavior of white females, some girls deftly manipulated public rhetoric about whiteness, heterosexuality, and womanhood to reinstate the due process lost to them, especially as white women, in the justice system. Constitutional protections to due process did not apply in mid-twentieth century North Carolina juvenile courts. Theoretically, the juvenile system endowed judges with almost unchecked discretion in determining what kind of discipline suited "the best interest of the child." In the reality of meting out justice, judges, prosecutors, and law enforcement took into account an adolescent's gender and race in treatment, sentencing, and incarceration. A chief mission for the juvenile courts rested in supervising whiteness in women. In the 1940s, courts and correctional authorities worried that correctional facilities, replete with the feebleminded poor, fundamentally threatened heterosexual norms that maintained the racial and gender order in society at large. Monitoring and supervising southern white womanhood, a category that served as a bulwark to heterosexual norms, became ever more important to the courts. Prison conditions rendered all prisoners susceptible to sexual immoralities, but white girls who engaged in illicit sexual liaisons required especially careful supervision and quarantine.

Specifically, this chapter traces the Samarcand offenders through Central Prison and contrasts the historical conditions that women and girls experienced in North Carolina's Central Prison and in the juvenile system that sustained women's and girls' reformatories. Armstrong had sentenced girls to the adult prison because girls and their guardians had persuaded him to believe that Samarcand Manor and the sexually illicit actions of the inmates there threatened their identities as white heterosexual females. As this chapter will show, conditions at Central Prison were pretty horrible. Nonetheless, adolescent girls, who understood the extreme restrictions, indefinite sentences, and intrusive and indefinite practices of parole at Samarcand, did their best to influence judges through reformers and ad litem guardians to convict them to adult prisons where, despite the hellhole conditions, sentences for misdemeanors were short and well-defined, and the parole system was far less intrusive. Adult female criminals served time in atrocious conditions, indeed. But imprisonment at white girls' and white women's reformatories such as Samarcand and the Industrial Farm Colony for Women meant that girls suffered the additional restrictions required of supervising whiteness in women, a project that the state had given up among its female offenders at Central Prison. In the case of the Samarcand teenagers, Central Prison discharged them after one year. At Samarcand, they would have remained until age twenty-one and possibly would have suffered continued indefinite commitment at the State Industrial Farm Colony for Women.

Samarcand had failed its mission as a bastion for reinforcing southern whiteness in women. It had become so damaging in the minds of southern whites that, among observers on the outside, even Central Prison in Raleigh seemed a better alternative for adolescent girls. Sexual perverts, mysterious policies, and lack of legal representation created a retrograde environment that transformed girls into more serious reprobates than they had been when they entered. Judge Armstrong abided by rumors that the institution is "not good," that many women and girls released from the place were embittered toward society and people in general, "and were worse off than before they were sent to the reformatory." For years he had struggled over similar decisions where women and girls preferred prison to Samarcand. Samarcand, he confessed, "has always been a mystery to me." Neither he nor the courts nor the people of the state really knew anything about the institution, he observed. No judge, he professed, should sentence a woman or girl to an institution and not know or care what becomes of her. "Once the gates are closed behind her," he noted, "she has very little opportunity to speak for herself and there are few to speak for her."[2]

In sentencing a child as an adult, Armstrong had turned the juvenile court system on its head. Since its inception in 1918, Samarcand had served as an important component of the juvenile justice system in North Carolina. Reformatories and juvenile courts represented the more humane alterative to adult prisons for juvenile offenders. Courts had the exclusive jurisdiction over children under age sixteen who were delinquent, dependent, or neglected, whose custody was in controversy, or who required the protection of the state. Once a juvenile court served summons on a child or parent, that jurisdiction attached to the child until age twenty-one unless the court transferred the child to an institution or ended the attachment by court order. Judges possessed considerable discretion over the court cases involving children. While adult due process involved proclamations of guilt or innocence, juvenile courts subordinated allegations of guilt in particular crimes to investigations of a child's character and the social forces that had contributed to the circumstances that led the child to court. From these investigations judges determined a course of action to save the child "in danger of becoming delinquent" or to save the child from destitution and neglect. Children escaped the death penalty and imprisonment with adults. But the state offered judges almost carte blanche powers, even in cases of dependency that did not involve a crime, in determining a child's fate and for how long. Thus, a more amorphous system of due process generally governed criminal procedure for children than for adults.[3]

This amorphous system applied to women's reformatories as well. Juvenile and women's reformatories radically upturned the traditional male imprisonment model. Both expanded the power of the state to police, penalize, and punish juvenile and female misdemeanants. Reformers, including Kate Burr Johnson, preferred indeterminate commitments of one to three years for female misdemeanants and argued that this form of treatment was more humane than prison. White women who committed felonies might land in Central Prison. But women convicted of lesser charges involving illicit sexual activity or drinking might find themselves sentenced to the North Carolina Industrial Farm Colony for Women in Kinston. The Farm Colony for Women, an achievement of middle-class women reformers, operated from 1929 to 1947. It served as a Samarcand Manor-style reformatory for adult women. Sometimes authorities transferred girls at Samarcand Manor to the farm colony when they turned eighteen. Treatment and punishment echoed the juvenile court system. Women lived on an isolated farm in a cottage-like atmosphere. Supervision was intensive and sentences were indeterminate. Whether or not a woman served at the farm colony largely depended on her community reputation and the degree of interest welfare workers and courts took in dealing with her situation. Family members, community workers, health officials, and court authorities usually filed complaints to have a women committed. Most of the women at the farm colony were under age forty and were daughters or wives of tenant farmers or cotton mill workers. The Farm Colony for Women simultaneously served the functions of a state-run asylum, a poor house, and a reformatory for adult women.[4]

Courts used these reformatories as reminders to white girls and women that the state would not tolerate the behavior of women who digressed from the norms of southern womanhood. Misdemeanants, defined as those girls and women engaged in vagrancy, truancy, sexual activity, drinking, fighting, and minor property crimes risked very long-term sentencing. Girls and women who landed in these reformatories served as reminders to young women back home that runaway behavior or the mere skipping of school had profound repercussions for their adolescent years. Parents, too, might threaten to send a daughter to Samarcand unless she should change her ways. This threat might reverberate deeply in the psyche of a girl, especially if she knew someone committed there who did not return. For a girl to claim the privileges of whiteness, she had best behave at home or else risk it all.

A child's gender and race mattered in the courtroom, too. Judges applied punishment and oversight very unevenly. A child endured specific punishments and oversight based upon his or her race and gender status. North Carolina operated 108 juvenile courts, including eight city juvenile courts, six combined city and county juvenile courts, and ninety-four county juvenile courts. In 1934, Wiley B. Sanders, a professor of social work at the University of North Carolina at Chapel Hill, and William C. Ezzell, a social work field agent, conducted a study of the juvenile courts from 1929 to 1934. Their intent was to study the incidence of delinquency, dependency, and neglect in children by gender and race to determine the court activity during the Great Depression years. Their conclusions found that the Depression years saw no significant increase in juvenile court activity or in types of crimes committed. However, they did discern significant variations in how individual judges applied the law, in courts' interpretations of certain vaguely defined charges such as "delinquent," "immorality," and "dependency," and in disposition of cases by gender and race. Their very comprehensive study of nearly all the courts in the state included 16,685 children: 52.8 percent "white," 47.1 percent "Negro," and 0.1 percent "Indian." Males accounted for 75.1 percent of the white children and 82.8 percent of the African American children. Courts processed most male cases as delinquent (over 85 percent). Courts far more often found girls (52.4 percent) dependent and neglected than boys. Sanders and Ezzell published two tables illustrating their data, reproduced in adapted forms in this chapter as tables 4 and 5.[5]

Boys accounted for the vast majority of delinquency cases in the state. Courts handled the cases of 6,575 white boys and 6,473 African American boys. The courts disproportionately charged boys by race. African American children accounted for 30.7 percent of juveniles, but made up 48.7 percent of those charged with juvenile delinquency. See table 4. Larceny was the most common crime among boys, though far more overrepresented among African Americans. As for treatment, courts usually placed boys on probation. However, white boys were more likely to land in an institution, such as the Stonewall Jackson Training School for Boys, than African American boys. Courts charged white boys more often than African Americans with charges of breaking and entering, injury to property, rape and attempted rape, drunkenness, "beyond

TABLE 4. Percent Distribution of Delinquency Cases, City and County Juvenile Courts of North Carolina, 1929–1934 (By Charge, for Each Race and Sex)

CHARGE	TOTAL	WHITE BOYS	WHITE GIRLS	NEGRO BOYS	NEGRO GIRLS
Larceny	38.5	34.7	13.5	46.8	29.6
Breaking and entering	10.3	12.7	0.5	11.0	1.4
Truancy	7.2	6.2	7.3	7.8	8.8
Delinquency	7.1	7.2	25.9	4.2	8.2
Fighting	5.6	4.1	4.3	5.5	15.7
Injury to property	4.1	6.5	0.7	2.9	0.5
Nuisance and disorderly	4.0	4.3	3.8	3.8	4.0
Beyond parental control	3.3	3.5	12.4	1.3	7.8
Assault deadly weapon	3.0	2.0	0.4	3.6	7.1
Runaway	2.0	2.7	4.3	0.8	2.9
Immorality	1.9	0.7	17.8	0.4	3.8
Trespass	1.7	1.8	0.7	1.9	0.8
Violation liquor law	1.2	1.6	0.5	1.0	1.0
Larceny of auto	1.0	1.5	0.1	0.7	0
Violation of probation	0.9	0.3	0.6	1.3	1.9
Violation of city ordinance	0.7	1.0	0.4	0.6	0.4
Drunkenness	0.7	1.3	0.5	0.2	0.8
Burglary	0.7	0.9	0	0.7	0
Carrying concealed weapon	0.6	0.4	0	0.8	0.2
Gambling	0.3	0.4	0	0.4	0.1
Forgery	0.3	0.4	0.2	0.2	0.2
Robbery	0.2	0.1	0	0.3	0
Rape	0.1	0.2	0	0.1	0
Attempted rape	0.1	0.2	0	0.1	0
Murder	0.1	0.1	0	0.2	0
Crime against nature	0.1	0.2	0	*	0
Manslaughter	0.1	0.1	0.1	0.1	0
Other burning	0.1	0.1	0	0.1	0
Arson	*	0.1	0	0.1	0
Incest	*	*	0.1	0	0
Miscellaneous	4.1	4.7	5.9	3.1	4.8
Total	100.0	100.0	100.0	100.0	100.0

Source: Table adapted from Wiley B. Sanders and William C. Ezzell, *Juvenile Court Cases in North Carolina, 1929–1934* (Raleigh, N.C.: State Board of Charities and Public Welfare, 1937), 19.

Note: The Samarcand arson defendants do not appear in this table because the General Assembly had stripped them of their juvenile status and their case was tried in the Moore County Superior Court. Also, Sanders and Ezzell denoted some categories by asterisk, though raw numbers existed.

parental control," "crimes against nature," and running away. Courts less often charged white boys with gun-related crimes, homicide, and larceny.[6]

A girl in juvenile court experienced very different treatment than a boy. Courts charged girls, both white and black, with "crimes" based on strict standards of behavior that did not apply to boys. In law, the long list of charges in the juvenile court system applied regardless of race and sex. Juvenile courts claimed jurisdiction in felony cases such as murder, arson, burglary, breaking and entering, and rape or incest for children under fourteen. Jurisdiction fell to the juvenile court in any misdemeanor case, such as truancy, running away, "nuisance and disorderly," "immorality," and "beyond parental control," committed by a child under sixteen. In practice, courts most often charged girls with crimes related to behavior. See table 4. White girls were three times more likely than white boys, and six times more likely than African American boys to be charged with "beyond parental control." White girls also suffered more often from the charge of "immorality." Courts applied "immorality" (1.9 percent of total cases) to 153 white girls (17.8 percent), 35 African American girls (3.8 percent), 38 white boys (0.7 percent), and 25 African American boys (0.4 percent). Sanders and Ezzell explained that courts regularly applied a gender-specific double standard in these cases. "The explanation for this difference in distribution between the sexes is probably that the girls of both races are held to a stricter standard of behavior than are the boys," they argued. The gendered double standard was so pervasive, they noted, "it is probably also the explanation for the excess of runaways among girls as compared to boys."[7]

In punishment, gender and race factored, too. Courts most often assigned a probation officer to supervise boys and African American girls in their home environments. Judges more often preferred to punish white girls (40.8 percent) by committing them to institutions, such as Samarcand Manor. See table 5. For more serious crimes, judges might send white boys to the Stonewall Jackson Training School and African American boys to Morrison Training School. No state reformatory existed for African American girls, whose cases judges usually referred to probation or eventually dismissed. But in the case of white girls, even minor cases of dependency or neglect, girls experienced institutionalization far more often than the total rate of commitment (16.9 percent) of all children. Judges varied wildly in their views on institutionalization, so that girls could not predict their case's outcome except by knowing a judge's reputation. "One juvenile court judge must be convinced a girl is 'immoral' before he

TABLE 5. Percent Distribution (According to Disposition) of Delinquency Cases, City and County Juvenile Courts of North Carolina, 1929–1934 (By Race and Sex)

DISPOSITION	TOTAL	WHITE BOYS	WHITE GIRLS	NEGRO BOYS	NEGRO GIRLS
Probation	38.4	38.9	30.0	38.2	44.6
Placed in institution	16.9	22.4	40.8	9.9	7.9
Dismissed	11.0	10.6	8.2	11.0	15.9
No disposition or pending	6.7	7.2	6.6	6.3	7.4
Jail or detention	5.6	3.2	.6	8.5	6.1
Whipped or punished by parents	4.5	2.1	.2	7.3	3.4
Restitution or damages	3.7	4.7	.1	3.6	1.8
Foster home or with relatives	3.6	2.3	4.4	4.9	2.4
Fine and cost	2.7	2.5	.2	3.0	4.0
County home or workhouse	1.6	.8	.2	2.5	1.5
Sent to higher court	.8	.7	.5	.9	.4
Returned home	.7	1.0	1.2	.5	.4
Other disposition	3.8	3.6	7.0	3.4	4.2
TOTAL	100.0	100.0	100.0	100.0	100.0

Source: Table adapted from Wiley B. Sanders and William C. Ezzell, *Juvenile Court Cases in North Carolina, 1929–1934* (Raleigh, N.C.: State Board of Charities and Public Welfare, 1937), 28.

commits her to Samarcand," Sanders and Ezzell observed, "while another juvenile court judge sends a girl to Samarcand to prevent her becoming immoral."[8]

The law prohibited courts from jailing children, but judges did so regardless. When girls landed in county jails, they experienced sexual assault almost from the moment they arrived. In a 1945 study of ninety-five delinquent children who spent time in jail, University of North Carolina-Charlotte social work student Ruth Thayer Hartman discovered that the children experienced filthy conditions, lack of privacy, little protection from adults, food deprivation, and sexual assault. Boys reported harassment. Several girls reported traumatization by sexual assault. One fourteen-year-old girl wrote:

In _____ county jail something awful happened there. A trustee does the cooking there. The jailor didn't run a very good jail. He let them talk just as they pleased. Mr. G.—[the jailor] had sexual relationships with a girl who was in there for murder, and his wife was down stairs sick. Three boys were across from me when I left. All three of them asked me if I would have sex relations with them. They asked me that and said they bet Mr. G—would have sex relations with me. All they talked about were things like that, babies and all, how they were born. I don't know much about birth.

One Samarcand girl who spent time in a county jail reported that she was once assaulted by the sheriff's son after an unsuccessful attempt by the sheriff to assault her. When the State Board of Public Welfare investigated, the sheriff resigned.[9]

African American girls were most at risk of incarceration in county jails because no state reformatory existed for them. Forms of discipline and treatment strongly reflected white supremacist beliefs that African Americans were inclined toward criminal behavior. The juvenile court system virtually ignored African American girls who misbehaved, as if magistrates and law enforcement expected that African American girls were naturally inclined toward causing trouble. Judges most often cited African American girls for larceny and "fighting," a charge courts applied three to four times more often to African American girls than they did to other groups. Unable to institutionalize them, judges placed 44.6 percent on probation, dismissed 15.9 percent, and committed only 7.9 percent to an institution, probably the Efland Home for Girls, a private institution run by women reformers. Judges were far more likely to fine African American girls, place them in adult jails or detention, or order them whipped.

Authorities argued that a juvenile court system inattentive to delinquency among African American girls failed to reinforce obedience and servility among them. J. G. Wooten, chief of police in Winston-Salem, advocated a court system more responsive to African American girl delinquents to reduce their criminal behavior as adults and stave off community costs of "thousands of dollars in the future." He complained that African American girl delinquents misbehaved because they knew the courts had no place to put them. Once warned and placed on probation, African American girls continued their delinquencies, placing "our community at the mercy of this particular group of girls," reformers complained, who are truants, steal, lie, and who are "immodest, use vile and obscene language," and, worse yet, are "lazy, impertinent, disobedient," and are spreading venereal disease by joining "several clubs for the purpose of prostitution." Their specialty, Wooten observed, "is grown men, their popular price per date is fifty-cents . . . THERE IS AN URGENT NEED FOR A HOME TO CARE FOR THESE GIRLS," he pleaded.[10]

Sheriff Wooten made this comment in 1934; the only reformatory for African American girls at the time was a private institution, Efland Home, the North Carolina Home for Colored Girls. Dr. Charlotte Hawkins Brown, president of the North Carolina Federation of Negro Women, had worked tirelessly since 1911 to establish the institution. She formed a biracial coalition with the North Carolina Federation of Women's Clubs and petitioned the legislature for a state reformatory for African American girls. In 1943 she appeared before the General Assembly to ask for money to support the struggling Efland Home. The legislature granted $50,000 for a public institution. That year, the State Training School for Negro Girls opened in Rocky Mount. In 1947, the school moved to Kinston and was renamed "Dobbs Farm," and then "Dobbs School for Girls." By 1949, twelve girls filled the reformatory to capacity and it received support from numerous white and African American women's groups. Although the school served far fewer girls than Samarcand Manor, this action by women's groups was a symbolic step toward challenging the conventions of whiteness that restricted ladyhood to white girls. Yet it had little practical effect in equalizing race relations, as the school remained segregated until the 1960s.[11]

Where juvenile courts provided no treatment to African American girls, they sentenced African American boys almost as they did adult male offenders. African American boys had the highest rate of delinquency (93.9 percent) and the lowest rate of dependency and neglect (5.5 percent) of any group. Larceny was the most common crime among boys, but it was disproportionately applied to African Americans (about 3:2). Most African American and white boys (80–90 percent) appeared before the juvenile courts only once, but African American boys were more likely to become repeat offenders, and a few appeared in court a dozen times or more. Sanders and Ezzell argued that the high rate of recidivism in African American boys likely was due to the low number of African American probation officers hired to supervise the offenders. African American boys also had the highest rate of whipping (7.9 percent). Twice the attorney general had ruled against whipping juveniles, yet nineteen courts whipped 589 children between 1929 and 1934. Eighty percent were African American. One city juvenile court judge frequently ordered parents to whip their children in front of the judge and police, usually with a paddle or switch.

This court had no probation officers and usually whipped multiple offenders. One boy required five men to hold him while he was whipped.[12]

African American boys (8.6 percent), more than any other underage group, landed in adult detention centers and in jails. Only three juvenile courts had separate detention facilities for juveniles. All others housed juveniles in the same building as adults. Ezzell and Sanders visited African American boys, some as young as nine years old, in county jails. One county housed all men, women, and juveniles in cells that shared a common corridor. The only privacy for women was a canvas sheet that partly draped over the front of the cell. Other counties held African American boys as young as twelve in dark, filthy city lock-up cells. There, boys freely associated with adult male offenders. Chief J. G. Wooten of Forsyth County attributed the high rate of recidivism in Winston-Salem to a bad environment caused by the closing of the Forsyth County Reformatory, which "turned loose on the streets of Winston-Salem and community, a crowd of delinquent colored boys, the products of broken homes and bad environment." The boys later turned up in the adult detention center. Wooten himself began a "City Juvenile Detention Home for Colored Boys" for the purpose of "trying to save the youthful potential criminal from becoming a criminal."[13]

In these patterns lie the contours of white supremacy in depression-era North Carolina. Juvenile court records and the discourse surrounding them demonstrate that for children, expectations of behavior varied according to race and gender. Courts expected conformity from youths who were not yet men or women, and whose varied skin tones categorized them as black or white. County authorities expected "white" girls to conform to ideals about purity and chastity. They arrested girls judged "immoral" and "beyond parental control," then institutionalized them until they succumbed to compliance. They judged dark-skinned girls, too, but society viewed this group as inherently disobedient, lazy, and promiscuous, irredeemable and unworthy of institutionalization. Courts viewed African American boys as more worthy of redemption, but more inclined toward violence and criminality, thus requiring more force and discipline to subdue. As for light-skinned boys, they were arrested less often. Courts looked askance at their general naughtiness and charged them only with the most serious offenses.

The juvenile court system and Samarcand Manor thus served as a vital social and legal institution that reinforced white male supremacy. The categories of whiteness and blackness, manhood and womanhood, were so restrictive as to require careful attention to the behavior of unsocialized youth, especially white girls. The juvenile court system thus marginalized anyone regardless of color or sex who challenged the ideals of white men's status as upstanding citizens, white women's roles as dependent and chaste, black women's servility and sexual availability, and black men's docility. Girls' and women's reformatories served the function of isolating and reeducating girls and women who failed to conform to stereotype. Cottage-style reformatories such as Samarcand Manor and the North Carolina Farm Colony for women appeared less punitive than men's prisons, but the "home-like" atmosphere subjected women to more intensive surveillance and control than men or women received in jail cells. Matrons and teachers held female prisoners to strict standards of behavior, language, and attire, and vigilantly monitored friendships for signs of homosexuality.[14]

In such a restrictive environment, how did a girl persuade a judge to break the rules? Perhaps she convinced the judge that the reformatory was a place not for white women but for sexual reprobates. Judge Armstrong's reference to rumors that "all sorts and forms of sexual perversion exist at this place" indicates just how low Samarcand Manor had sunk in the minds of reformers and state authorities. It was no mere slip of the tongue. His concerns reflected changing historical constructions in the 1930s and 1940s of homosexuality and deviant behavior. Until 1930, many social scientists had attributed same-sex sexual encounters in prison as situational behavior or "pseudohomosexuality" that would not have occurred in open society. Generally, the image of aggressive lesbian homosexuals accompanied racial connotations. Americans typically viewed white women inmates as heterosexual prostitutes who engaged temporarily as lovers to aggressively lesbian African American women. In the 1930s the popularity of eugenics shaped new interpretations of same-sex relations. Eugenicists and mental hygienists increasingly attributed same-sex sexuality as "abnormal" or "pathological" deviance associated with feeblemindedness among the poor in both races. By the 1940s and 1950s, sex crime panics and investigations of homosexuals in government represented new Cold War-era anxieties about homosexuals as threats to the nation's security and children. Prison life, where sexual identities separated "wolves," "punks," and "pansies," fueled fears that society at large might dissolve into a nation of homosexual psychopaths. The 1940s thus witnessed new cultural constructions of heterosexuality as "normal" and "natural" and prompted new fears of homosexuality and homosexual identity that had not previously existed. New language, such as the word "deviant" to describe prison homosexuals, contributed to the process of "unnaturalizing" homosexuality and of "naturalizing" heterosexuality in society at large.[15]

Delinquent white girls in boarding schools and reformatories generally escaped most of the scrutiny spurred by anxieties of same-sex encounters. Before Armstrong issued his alarm, reports, letters, depositions, and newspaper accounts referenced very few incidents related to sexual improprieties at Samarcand. Occasional reference exists to masturbation among girls. For example, after the arson trial, staff and social workers discussed sexual practices at Samarcand. Some former staff members had complained that girls practiced sodomy and masturbation. Agnes MacNaughton conceded that the reports were "probable," and that the behaviors were difficult to control among a population of girls with "sex experience." Sometimes county social workers complained of rumors of sodomy and "sex perversion" practiced at the school. These rumors stoked fears in the general population when Samarcand discharged girls back to their communities. But staff usually described these incidents as isolated and generally reflective of immoral and antisocial behavior. Usually staff avoided direct references to sexual behavior among the girls. Nationwide, prison staff referenced same-sex sexuality by refusing to name it, or used a "language of elision" to vaguely reference indecencies too horrible to refer to by name. Prison officials were acutely aware that women and girls engaged in same-sex sexual encounters. However, experts usually dismissed these encounters as situational schoolgirl friendships. One psychiatrist, Maurice Chideckel, noted in 1938 that female inmates engaged in sexual encounters but usually retained their femininity. In reference to prostitutes, he said, "as soon as they are let out of the institution they resume business at the old stand." Other women, experts insisted, turned to same-sex encounters after traumatizing pre-prison experiences with men.[16]

Armstrong's comments suggest changing views in North Carolina of sexuality and the behavior of delinquent white girls. In 1943, North Carolinians no longer dismissed improprieties at Samarcand as schoolgirl friendships. His direct reference to rumors of "sexual perversions" strongly indicates that he feared girls at Samarcand might reenter society not as southern ladies but as mannish, weak-minded lesbians. In southern white minds, Samarcand no longer represented a school that reformed bad white girls into good southern ladies. The language of "perversion" placed Samarcand inmates in the same category as offenders in adult prisons, where homosexual deviance indicated mental weakness. It is likely he worried that white adolescent girls, already vulnerable to deviant behavior, would experience same-sex sexual relations and embrace homosexuality. This context is important for fully understanding his comments about the environment at Samarcand and the rumors that the reformatory had violated its mission to transform unruly white girls into genteel southern ladies.

Armstrong clearly believed that white girls were better off at the penitentiary. But were conditions at Central Prison better or worse than the girls' reformatory environment? To answer this question, one must know the horrible conditions that historically plagued women in penitentiaries. Pity a woman in a men's prison. Neglected, vulnerable to assault, and always subjected to harsher punishment than men, she is a pitiable creature indeed. Scholars have shown that nationwide, women locked up in men's prisons consistently suffered a gendered double standard. At the very least, wardens ignored them, disregarded their health needs as women, and relegated them to inferior conditions. Nicole Hahn Rafter has referred to women's treatment in men's prisons as "partial justice." Prison administrators, by their neglect, excused women from rules and assigned them to less physically demanding labor. However, administrators' partiality did not produce better treatment. Women had fewer opportunities for fresh air, had less surveillance and less protection, suffered from loneliness more often, and felt more stigmatized than men. Women also felt humiliated by lack of privacy from the opposite sex, particularly by male wardens, and by threats of forced prostitution and rape. Child-related problems plagued pregnant women. Women might be locked in cells identical to men's, but they had to deliver their infants and try to keep them alive.[17]

Contemporary accounts serve as direct witness to these conditions. In 1936, Mary Belle Harris, who served as a warden at a predominantly male prison, The Workhouse at Blackwell's Island in New York City, noted the neglect and idleness that women prisoners suffered. Men, she argued, performed all of the outdoor work, while a few women labored inside at hot and confining tasks, such as sewing, laundering, and housekeeping. Most women had no work at all, she reported. Day after day, month after month, most women sat just watching, waiting, and talking. "It is difficult to give a conception of the place to anyone who has not seen it—grim and gray, filled with miasmas, physical and mental, with nothing to counteract and arrest the deterioration which inevitably results from idleness under such conditions."[18]

Neglect bred disease and criminality. Kate Richard O'Hare, a socialist agitator convicted at age fifteen under the Espionage Act of 1917, served time in the state prison for men in Jefferson City, Missouri, from April 14, 1919, until President Woodrow Wilson commuted her sentence in May 1920. The women's cell house, constructed of stone and concrete with tile and cement floors, housed about one hundred women. The building was better than some of the men's cell houses, but "less satisfactory" than the most modern prison buildings for men. It had plentiful windows, but they were filthy and painted gray to shut out the light, and it had a "woefully dilapidated heating system." Crude furniture, including a steel bunk and straw mattresses, adorned each cell, and "every crack and crevice of the cellhouse was full of vermin of every known sort." The prison, she complained, made no effort to segregate the young offenders from the more hardened and vicious criminals. But by far the worst, most loathsome and inexcusable conditions involved the lack of treatment for communicable diseases. No hospital, no sick care, no medical equipment or supplies, no quarantining of the sick existed in the women's wing. Tubercular and syphilitic women prepared food and shared bathing facilities with uninfected women. O'Hare describes women bathers with masses of open sores dripping pus in bathtubs and women food preparers with pus oozing from open sores on their arms and dripping into dishes. During an influenza epidemic, administrators locked forty or more women, including O'Hare, in their cells with no care at all. No effort was made to segregate or care for the mentally diseased. "It is a tragic and soul-sickening thing," O'Hare remembers regarding treatment of mentally diseased women, "that the most revolting instances of brutality and downright fiendish cruelty were directed toward the women utterly unable to make the 'task' or conform to required discipline."[19]

Women received worse treatment than men because men's prisons stripped them of their femininity and protections associated with womanhood. Women criminals entered prison as "fallen women," "whores," or "morons" who no longer deserved the dignity associated with the protections of white womanhood. In prison, their womanhood exposed them to gender-specific physical and mental violence. Authorities treated them not just as criminals but as subhumans who required neither attention nor civility. Even women convicted of misdemeanors risked losing their gender-based protections and castigation as "fallen women." Prison conditions for women had changed little in a century. One New York chaplain in 1830 captured the feeling of this sociological misogyny when he commented that "to be a *male* convict in this prison would be quite tolerable, but to be a *female* convict . . . worse than death."[20]

Social scientists commonly ascribed gendered characteristics that differentiated between the male criminal mind and its more subhuman female counterpart. The famous Italian criminologist Cesare Lombroso perpetuated these attitudes in the publication of a forty-year study of the "female criminal mind." He concluded that women criminals were "big children," "more dangerous," and possessed more evil tendencies than men. In 1928, a Chicago judge, Marcus Kavanagh, agreed. In *The Criminal and His Allies,* Kavanagh wrote that "no man can be so bad as a wicked woman. He hasn't the same genius for evil. A woman is always more hurt by her fall than is a man by his fall, for the reason that a man drops from the first-story window, while a woman tumbles from the roof." He blamed "women's liberation" and women's "entry into the wider world" for enlarging the number of women prisoners.[21]

The experience of the twelve Samarcand inmates convicted of arson in 1931 and sent to the Central Prison in Raleigh indicates that women suffered similar neglect, isolation, and endangerment. In North Carolina, few records exist as to the fate of women prisoners in the state's Central Prison. However, a careful reading of administrators' reports, such as the *Biennial Report of the Superintendent of the State Prison Department, 1930–1932,* and internal letters, such as the governor's correspondence, suggest that North Carolina's female inmates suffered similar circumstances as women elsewhere in the nation. Women were among only a small minority of inmates in the men's prison. In 1932, Central Prison's population of 635 inmates exceeded twice the maximum capacity of the buildings. Of that total, 111 (38 white and 73 "colored") were women. Most women were convicted for crimes of arson, homicide, larceny, theft, and burglary. Superintendent George Ross Pou noted the dangerous conditions suffered by all inmates. The whole prison, he said, "represents the very worst type of *fire trap* imaginable." Furthermore, despite the maintenance of a "Women's Ward," the prison lacked any policy to segregate by race, sex, age, or type of offender. Finally, no policy existed to quarantine prisoners with communicable diseases.[22]

In prison, the Samarcand arson teenagers lived in proximity to the very worst adult male offenders. When the girls first arrived at the prison in March 1931, the warden imprisoned them in the men's wing, in small cells above death row. The warden, mindful of their history of burning dormitories and county jails, placed them in the only fireproof building at Central Prison. These small cells, known as "regulation" or "isolation" cells, had concrete floors, were inadequately ventilated, and were practically dark. Only a few slats in the solid wood doors allowed any light or air into the cell. The girls likely spent almost every moment in these cells without exercise or socialization. Only by this extreme solitary confinement could the prison staff minimize the girls' interactions with the state's most felonious male offenders. Even so, the girls encountered some interaction with men. Superintendent Pou noted that after several weeks or months in isolation he ordered the girls moved to the Women's Ward because his staff had failed "to bring about the desired segregation of male and female prisoners."[23]

Yet conditions in the Women's Ward were also deplorable and, furthermore, known as dangerous. When State Insurance Commissioner Dan C. Boney learned that the convicted arsonists had arrived in the Women's Ward, he threatened to cancel the insurance coverage on the building. The Women's Ward was among the oldest buildings at the prison. Already at twice its maximum capacity, the building was one of two wings constructed in the 1870s entirely of longleaf pine cut in the late 1860s. Pou pleaded for Boney to maintain coverage and promised to take the matter to the governor. "It seems a crime for the State of North Carolina," Pou informed Boney, "to have to confine these girls in small cells continuously, as we did for the first few weeks after their commitment to the State Prison." The warden could not keep them in the Women's Ward. To lose fire insurance on a building when there was a daily danger of fire was poor policy and, at the very least, would not look good to the press. Meanwhile, Pou informed Governor O. Max Gardner of the liability. "You are well acquainted with the fire hazards at Central Prison," he noted, "where 638 prisoners are confined. . . . I shudder when I pause to think what might occur should there be a fire in either of these wings." Pou recommended that the governor remove the women prisoners to a camp in Cary, North Carolina, where they would work in prison sewing shops and the laundry.[24]

Endangerment of prisoners continued. A crisis sparked controversy in late 1935 that caused the legislature to authorize funds for a women's prison. On September 15, 1935, the sheriff of Laurinburg, North Carolina, dumped two African American female prisoners sentenced to road work at a public works camp that had no housing facility for women. In the 1930s, North Carolina outlawed the contract lease system of prison labor in mines and quarries and instead shifted to state-run industries that concentrated prisoners in highway construction, prison laundry services, and license plate construction. These "chain gang" highway prisoners came under the authority of the Highway and Public Works Commission (HPWC). The HPWC staff had no housing for the women prisoners. With no place to put the African American women prisoners, the HPWC staff simply released them. Apparently, the two women were known as "notorious characters [and] a nuisance to the community." The incident created such a firestorm of debate about jurisdiction, procedure, and corrections authority that it finally forced the legislature to act. Late that year the North Carolina General Assembly directed funds to the building of a "Women's Industrial Prison" south of Raleigh.[25]

Although incarcerated in a predominantly male facility, healthcare for women existed at Central Prison. In the Women's Ward, the Samarcand girls suffered from a marginal medical system that promoted sterilization at the physician's discretion. A prison physician, Dr. J. H. Norman Jr., provided medical services at the prison hospital, and a prison surgeon, Dr. Kemp Neal, performed medical operations. Some of the Samarcand inmates likely sought out their care. The physician provided some attention to women's reproductive health, but little is known about the type of care. From 1930 to 1932, his reports stipulate that he saw five pregnant women and administered four live births and one abortion, but they provide no further information about the welfare of the children. He diagnosed two cases of amenorrhea and 109 of dysmenorrhea. The surgeon, Dr. Neal, performed four hysterectomies and in each case also removed one or both ovaries and fallopian tubes. In two additional cases he removed ovarian cysts or diseased tubes. Every case involving ovaries and hysterectomies effectively sterilized the patient. There is no indication that this physician operated under the orders of the Eugenics Board or a classifications board such as the one at Samarcand Manor. However, in four of these six cases he also performed an appendectomy, a procedure that some sterilization victims have complained served as a physician's justification for performing involuntary sterilizations outside of the Eugenics Board's scrutiny. In fact, an appendectomy where the physician also sterilized the patient was known as a "Mississippi appendectomy," a term used to describe nonconsensual sterilization performed on women inmates at the discretion of the physician. In the same period that he performed six surgeries on female genitalia, he also performed three surgeries on men. In one male prisoner he removed a "tube" in a penis; in another he performed an "operation on penis." And in a third he reported a "penis amputated."[26]

The Samarcand convicts bore their brutal sentencing for one year. Though imprisoned for terms of five to eight years, they benefitted from the newly instituted parole system and an extremely overcrowded and growing prison population. The prison simply could not afford to accommodate the firebugs, and they were released back into the general population. All twelve had entered the prison on May 20, 1931. The warden discharged them individually in the summer of 1932. Thelma Council, the first girl discharged, walked out on June 28. Josephine French stayed the longest. She left on August 27. The commutation books do not indicate where the girls resettled. It only indicates their county of birth and their last place of residence before incarceration at Samarcand. Presumably the girls relocated to their home counties, but this information is not available in the parole records. Only one girl, Ollie Harding, ever reappeared in the Central Prison register. In October 1934 she reentered Central Prison, convicted on charges of fornication, adultery, and vagrancy in Beaufort County, her place of birth. She served six months and was paroled on May 16, 1935. After 1935 the Samarcand girls disappear.[27]

North Carolina authorities did not want its naughtiest white girls to vanish into the general population. The imperfect parole system, designed for men, did not keep close track of women who might marry, change names, and move elsewhere. Ex-convict white girls who moved freely in their communities violated every precondition of proper white ladyhood. The only women who moved about the streets were prostitutes that the law could not catch and African American women and girls that the law ignored. Authorities feared the confusion that surfaced when white women turned to the streets. A street woman was, by definition, feebleminded, promiscuous, or both. She most certainly was not a white lady. Surely to the consternation of authorities, the Samarcand arson convicts had eluded the reformatory system. Even Judge Armstrong hoped to clear Samarcand's reputation so he could resume sending girls to the reformatories, as the law required.

To set the record straight, Judge Frank M. Armstrong ordered an investigation of Samarcand Manor. At the opening of the Moore County criminal court's January 1943 term, Armstrong instructed the grand jury to investigate the conditions at Samarcand. He appointed a committee consisting of powerful men in the county. Chairman G. C. Seymour of Aberdeen served in county politics and had founded the Coca-Cola Bottling Company of Aberdeen. James W. Tufts was a descendant of the founder of the Pinehurst golf resort. L. E. Pender of Southern Pines was their fellow Kiwanis member. Armstrong's decision to intervene in part rested on his own self-preservation. After all, he had to justify why he had violated his duties under the Child Welfare Act by sentencing children as adults. By sentencing children as adults, he had denied girls the protection of the juvenile system as mandated by law. His instructions to the grand jury, he explained, "were not intended to suggest that anyone had violated a law or to imply incompetence of those in charge at Samarcand." However, rumor abounded. He declared that his suspicions drew from "information" and "personally observed circumstances." It was in the interest of the judge and within the scope of his power to rest the public mind by investigating what had grown into widespread criticism of the girls' institution. To be sure, rumor did not carry weight as evidence in a courtroom. But Judge Armstrong's authority rested upon the doctrine of judicial discretion, a vital component of decision-making in the juvenile court, where judges were instructed to rule "in the best interests of the child."

Administrators, social workers, and even the governor jumped to defend Samarcand Manor against Armstrong's charges of sexual perversion. They strongly defended the institution as a refuge for young white girls victimized by sexual assault and excoriated Armstrong for insinuating the place was a harbor for prostitutes and lesbians. Women reformers demanded that female social workers serve on the grand jury committee. One manager of the Samarcand Manor board of trustees, George Herr, wrote a letter to underscore the importance of Samarcand as a reformatory for girls, not as a prison for seasoned criminals. The indefinite sentence provided a girl with time to gain a new start. Girls committed to Samarcand served indeterminate sentences and were released upon parole, violation of which would send the girl back to the institution. "It is not a matter of surprise to me that a confirmed prostitute, even though she be only sixteen or seventeen years of age," he argued, "would prefer a sentence to state prison. This sentence would be for a definite period and when completed she would be released unconditionally to continue to ply her trade." Governor Broughton agreed, and specified that Samarcand was an institution for girls under sixteen. State law, he said, forbade judges from committing to Samarcand adult women charged with prostitution. According to

Broughton, the institution recently had begun to admit girls as old as eighteen on prostitution charges "due to war conditions," but that generally the institution admitted only girls under age sixteen.[28]

Samarcand Manor staff, trustees, and the governor heartily defended Samarcand. When Armstrong instructed the grand jury to interview at least three inmates in every classification and age group, Broughton attacked Armstrong for overstepping his powers. The law rested investigation with the state, not county court judges, he said. Superintendent Grace M. Robson and W. A. Stanbury appeared before a separate legislative committee to defend the institution. It was upstanding, they said, but beleaguered. They pleaded for additional funding for better clothing for the girls. State appropriations had allocated only $4 for clothes per girl, each of whom was therefore obligated to wear substandard shoes and patched overalls. Mrs. Wilbur Bunn, another trustee, argued that the law required Samarcand to reintroduce the girls into their communities as "self-respecting women." The women at Samarcand, she argued, "are like other women—they have the same desire to make themselves as attractive as possible." The state could not expect to rehabilitate them "to occupy higher planes of society" without appropriate clothing.[29]

Those closest to the institution jumped to defend the reformatory as a reputable training school, a place to restore girls to the standards of white womanhood. Board members, the superintendent, reformers, and social workers embraced the indeterminate sentence and saw it not as a punishment, but as an opportunity for girls to rediscover their femininity and find their place in the racial social order. Adult prostitutes sent there also were denied due process and suffered the indeterminate penalty. Authorities, blinded by their devotion to racial purity, viewed the indeterminate sentence as necessary to control the feebleminded, a category that wayward girls and prostitutes met by definition. The constitutional rights of due process did not seem to apply when the defendant in question was a white woman or girl who failed to mature out of the wayward, incorrigible, and sexually promiscuous years of adolescence. In a case where a girl's mental sanity appeared borderline, a simple intelligence quotient test would diagnose her mental age in childhood years. For those who remained always "juvenile," due process rights were irrelevant.

Within weeks, Armstrong's investigation concluded. The grand jury report commended Samarcand Manor for its "healthy" environment. The grand jury noted the white linen dining facility staffed by "well trained and experienced dieticians," the wholesome cottages with "sleeping porches," the trained staff, systematic classification system, honor girl privileges, and a disciplinary system organized by a group honor system where the whole group suffered demerits or lost honor status whenever one member of the group committed an offense. The committee interviewed several girls, none of whom criticized the system. Two problems, the report concluded, most likely had perpetuated the rumors among adolescents that prison was better than Samarcand. First, the institution still suffered from the stigma of the bad reputation it had gained under the leadership of Agnes MacNaughton. Second, girls disliked the indeterminate sentence and invasive and indefinite parole system that allowed for recommitment to the institution. The board usually paroled girls after fifteen to eighteen months. Some of the girls, however, "who are more backwards are required to say [sic] longer to complete their training." The committee acknowledged that Samarcand required more parole officers and a full-time psychologist because "a large proportion of the girls are mentally subnormal."[30]

The report restored the court's faith in the institution. Samarcand, it concluded, was an appropriate institution for feebleminded girls who had no other advantage in society. The grand jury concluded that the long-lasting stigma gained during the arson trial was unfounded. Subnormal girls lived there, the committee found, but the state treated them well. Systematic procedures, a hospital, and a disciplinary system monitored by the girls themselves worked well to maintain order and fairness at the institution. None of the girls, it reported, complained at any length. Thus, the committee concluded, they were satisfied with their treatment. The juvenile court system remained intact and courts should continue to employ it in their arsenal of weapons against the tide of delinquency among girls. Although the report did not specify, its authors appeared satisfied by Samarcand's postwar function. That is, it operated as an asylum in the guise of a girls' reform school.

More important, the report allayed suspicions of indecent sexual activity at the institution. Samarcand Manor housed girls ages ten to sixteen. Older teenage girls committed for prostitution were admitted temporarily, they said. The grand jury committee did not perceive the older girls as a problem or threat to the younger ones. "We found that the opportunities for immorality or for sex abuse it [sic] at a minimum," the report concluded, "it being almost impossible for such to be engaged in on account of the close supervision which each girl has at all times." Thus, the committee concluded that Samarcand Manor was a wholesome and positive facility for the state's wayward white girls. "We were generally impressed with the business-like and systematic operation of Samarcand as one of the correctional institutions of state," they concluded, "and we believe the operation of this institution is such that it merits the continued support and cooperation of the state."[31]

Once the reformatory gates closed behind a girl, there was no telling what might happen to her. Stringent conditions, indefinite sentences, the reality of forced sterilization, floggings, and other cruel forms of punishments beset girls and women in the reform institutions. Misdemeanors such as vagrancy that would bring a white boy or man a light sentence would land a white girl or woman in a reformatory for an indefinite term and possibly end with her forced sterilization. The 1931 arson defendants had voiced their protest to some of these concerns when they said they had set the fire to escape the institution for jail. To be sure, prison conditions for women left much to be desired. Cells were crowded and in disrepair. Few programs existed for women's rehabilitation. Where girls in reformatories endured scrutiny by the Eugenics Board, women in the state prison were sterilized at the discretion of the physician. However, the male standard of classification, labor in laundries and tailoring, definite sentences, and unlikelihood of sterilization seemed preferable to some girls and women who faced a choice between a rock and a hard place.

The state's efforts at maintaining whiteness through racial purity had created a reformatory system that promoted a racial and gender hierarchy best described as a multitiered double standard. Painfully burdensome on the state's white girls, it enforced chastity; neglectful of the state's African American girls, it defined them as beyond reform and lost to society. White boys suffered reform in serious offenses, and African American boys suffered prison sentences for minor crimes. For juveniles in North Carolina, personal freedoms were not protected and due process did not exist. The threat of the reformatory and sterilization reminded white girls of every class that they had the most personal freedom to lose. That is why, in 1943, adolescent girls and their ad litem guardians very well may have viewed the state penitentiary as preferable to Samarcand.

Epilogue

IN 2011, NORTH CAROLINA CLOSED Samarcand Youth Development Center. Plagued by scandal and decimated by budget cuts, the old facility no longer met the needs of the juvenile justice system. New philosophies about recidivism have come to shape juvenile justice in the state and nation. In 1998, North Carolina embraced a new juvenile justice model comparable to that posed by Birdie Dunn in 1917 and began a slow dismantling of its reformatory system. In 2013, the North Carolina General Assembly allocated more than $10 million for a two-year renovation of the campus into a "state of the art training center" for the North Carolina Department of Public Safety. When complete, the old reformatory will have found a brand new purpose.[1]

Birdie Dunn, the prescient public health nurse who had opposed the 1917 legislation that founded Samarcand Manor, had predicted that no good would come out of a white girls' reformatory. She had argued for community-led programs and local supervision instead of the state centralized reformatory model for the treatment of white female delinquents. In 1917, advocates of the centralized model prevailed, but over the scope of nearly a century Samarcand and the reformatory system it represented became mired in reports of scandal, torturous physical and sexual abuse, and prostitution. This book has demonstrated the history of that reformatory model, rooted deeply in white supremacist and eugenics ideology. But even to the present day, as eugenics language has faded from state policy and the reformatories have integrated in response to the Civil Rights Act of 1964, formalized reports of abuse have persisted—not only at Samarcand, but at other reformatories in the state and nation. Evidence indicates that forced sterilization, beatings, inadequate and high turnover among staff, lax security, sexual abuse so rampant that it took on a culture of "normalcy," prostitution, illicit drug use, and even homicide have characterized the reformatory experience at the state's training schools and especially at Samarcand.

In the first decade of the twenty-first century, public data and personal recollections brought to the surface a culture of child abuse at reformatories throughout North Carolina. This book has discussed at length the whippings that occurred at Samarcand in the 1930s. But boys endured beatings as well. At Stonewall Jackson Training School, founded in 1909 as a reformatory for white boys, boys and parents complained that abuse was commonplace. In 1934, one father took his son, George White Goodman, to an attorney to pursue legal action for the beatings his son endured during his 1929–34 commitment. The attorney, Horace M. DuBose, observed that at their first office visit "the lower part of [the boy's] body was one solid mass of bruised flesh." Another boy testified that he had observed Goodman's cottage parent and school officer, Alfred Carriker, whip Goodman brutally and in rapid succession for three to five minutes with "a hickory stick about the size of a man's larger finger." Goodman received a terrible beating because Carriker had found stones in the beans which Goodman was cooking. "I know that Carriker was in the habit of whipping Goodman every day or so," James Cooper testified. "He seemed to take a delight in doing so, and always appeared in a happier frame of mind afterwards than before." The governor, J.C.B. Ehringhaus, called for an investigation by the State Board of Charities and Public Welfare. The board recommended minor reforms in parole and supervision practices, and called for record-keeping in disciplinary procedures.[2]

The reforms had little effect. Unchecked beatings occurred at Jackson well into the 1940s, 1950s, and 1960s. An expose in the *Charlotte (N.C.) Observer* by journalist Elizabeth Leland, published in 2013, details the horrors that former inmates, now aged sixty to eighty-three years old, remember from their brutal experiences as young boys at Jackson Training School. John Dollard of Asheville was sent to the school in 1964. He was charged with no crime, but landed in the reformatory because, he explains, he had "an alcoholic father who couldn't get over World War II." He describes the sadism that existed at the boy's school. The worst beatings occurred when boys attempted to run away. Upon capture, staff forced them to lie naked across a bench. "Three grown men took their time to beat you so bad you almost passed out," Dollard said, "it was beyond pain." Jerry Moore of Black Mountain reported that the staff tormented him by hitting him in the face, kicking him in the ribs, and slapping his penis with a rubber strap. The trauma lingers decades later. John Pate, locked up not for a crime but on the advice of a neighbor who thought he would benefit from the reformatory as opposed to living in poverty, explained that at age eighty-three the memories remain painful. "People who have never seen blood dripping off the toes of children will never understand what we feel," he said.[3]

"Stonewall was a nightmare," explained James Tompkins, former director of the Child Advocacy Commission, "it was out of the twilight zone." In the early 1970s, North Carolina ranked first in the number of children sent to reformatories per capita. In 1972, the North Carolina Bar conducted a study of all eight North Carolina reformatories and concluded that the state used the centers as a "dumping ground for unfortunate children." In 1974, after the shooting and killing of a runaway girl from Samarcand, Tompkins conducted an investigation of five of the reformatories and found extensive evidence of systematic abuse, sexual assault, prostitution, and illicit drug use. He compiled his data while living in cottages with children for three to ten days at each institution. He slept and ate in some of the cottages, accompanied children and adults on daily activities, and interviewed three hundred children and over one hundred adults at the facilities.[4]

His report, "Hell Without Fire: A Report on Children," stoked controversy among officials in the Department of Corrections, who characterized his findings as "outdated." It resulted in his dismissal and eventual reemployment as an instructor at Appalachian State University. His reports mainly focused on the corruption and abuse by staff. "Personnel who commit felonies and attack children are still maintained as employees," he reported. "Conditions are deplorable and without hope." Even in instances of repeated sexual abuse, predatory adults acted with impunity. Adults seduced both boys and girls at local motels and in their own housing. Older children preyed on younger children. Children from all institutions reported to him a constant, daily threat of "propositions and sexual overtures." "New children are propositioned frequently," he reported, "and are forced to submit." Tompkins reported incidents described by both adults and children as "animalistic, aggressive, torturous, and primitive sexuality." His very negative report depicted brothel-like activities at Samarcand.[5]

Reports of sexual abuse plagued Samarcand into the twenty-first century. The center, integrated in the 1960s, began accepting boys in the 1970s and since has served as a coed institution or a single sex center throughout its recent history. In the early 2000s, authorities redesignated it as a girls' institution. A federal study completed by the U.S. Department of Justice in 2009 ranked Samarcand high on a list of youth correctional institutions reporting sex abuse. In a nationwide survey of nine thousand youths in custody, Samarcand ranked among thirteen facilities listed as having a "high rate of sexual victimization." According to the study, sexual victimization is "any forced sexual activity with another youth (nonconsensual sexual acts and other sexual contacts) and all sexual activity with facility staff (staff sexual misconduct and staff sexual misconduct excluding touching)." Of Samarcand respondents, one-third reported sexual misconduct, exceeding the national average of 12 percent. Most incidents took place on the premises between 6 P.M. and midnight and occurred in a common area, office, shower, or sleeping area.[6]

The study stoked debate about the data collected from youth. Bart Lubow, director of the Annie E. Casey Foundation, an advocacy group for children, called the rates of misconduct "so high they're stunning." Linda W. Hayes, former secretary of the Department of Juvenile Justice and Delinquency Prevention, questioned the accuracy of the survey. "We have zero tolerance for any level of mistreatment of juveniles within our care," she argued, "any number above zero is unacceptable." Hayes reported that the survey was conducted in a time frame when the department documented a pattern of false allegations at the facility, and thus data may be "suspect." Allen Beck, a senior statistical advisor at the U.S. Department of Justice, acknowledged that the survey results at Samarcand are allegations. However, he maintained that "even if only half the girls there are telling the truth, which I'm not willing to concede, then they still have a serious problem."[7]

Nationwide, the 2000s witnessed a growing concern and awareness about reformatory abuses. In 2011, the Annie E. Casey Foundation published an issue brief, *No Place for Kids: The Case for Reducing Juvenile Incarceration.* The authors contended that youth incarceration does not reduce recidivism and only aggravates violence and abuse against children. In a survey of confined youth published in April 2010, 42 percent of youth surveyed "said they were somewhat or very afraid of being physically attacked, 45 percent said that staff use force against youth when they don't need to, and 30 percent said that staff place youth in solitary confinement as a form of discipline." Archaeological evidence bore the worst-case scenario. In January 2014, authorities discovered fifty-five bodies in unmarked graves on the grounds of the former Florida

School for Boys in Marianne, Florida. Researchers suspected that some of the boys at the school were killed by staff. The school, which closed in 2011, was notorious for its abuse. In 2009, more than two hundred former students filed abuse charges in a class-action lawsuit against Troy Tidwell, a former instructor at the school. The governor's office ordered an investigation, but the state attorney, Glenn Hess, declined to open a criminal investigation.[8]

In the 1990s, new juvenile justice philosophies replaced the youth detention center model to treat all but the most serious juvenile offenders. Evidence-based data on recidivism, new knowledge about adolescence and brain development, greater awareness of due process rights, and emphasis on individualized counseling, therapy, and education to reduce recidivism generated national and statewide restructuring in delinquency prevention, with an emphasis on reinvestment into community resources.

Nationwide, "rehabilitation and reinvestment" replaced the prison and reformatory models in both adult and juvenile offenses. Recent criminal justice policy scholars have argued that incarceration rates reflect public policy, not changes in the crime rate. In the 1970s, 1980s, and 1990s, incarceration rates soared as states adopted punitive lock-up policies. Localities, strapped for cash and unable to invest in county jails and community programs, sent prisoners to state corrections centers so as to shift the economic burden of prison care from the community to state budgets. In the 1990s several states could no longer financially maintain these large state prison populations and began to seek other means of prisoner care.[9]

Juvenile justice authorities adopted a new philosophy toward care of delinquents. California, Pennsylvania, and Wisconsin had pioneered innovative policies in the 1960s and 1970s to reduce the demand for expensive state confinement and to supervise young offenders in their own communities. In 1974, the federal government created the Office of Juvenile Justice and Delinquency Prevention to focus not only on treatment, but also on prevention programs, such as boys and girls clubs, that targeted youth at high risk for arrest. By the 1990s numerous states, including North Carolina, Ohio, and Oregon, began to follow suit. Juvenile justice reflected a philosophical shift: ensuring public safety and altering the life trajectories of juveniles to reduce criminality and improve productivity as citizens. Criminal justice experts accepted a new way of thinking: that institutionalization places youth in an organizational subculture of confinement that breeds violence rather than suppresses it. New "rehabilitation, reinvestment and realignment" models encouraged both juvenile and adult justice programs to shift funds away from confinement and incarceration and toward drug treatment, counseling and therapy, education, and community supervision for the purpose of facilitating positive behavioral changes. Community reinvestment also appeared more cost-effective. The prison model required budgeting of more than $100,000 per year per inmate. Community reinvestment carried positive returns. A 2011 Texas study demonstrated that from 2006–2011, a $100 million reinvestment in community-based programs dropped the "youth inmate" population more than 60 percent.[10]

In 1998, North Carolina adopted the Juvenile Justice Reform Act, a "rehabilitation and reinvestment" model, to reflect these new philosophies and cost-saving structures. North Carolina is one of two states that caps juvenile status at age fifteen. Youths aged sixteen to eighteen are charged as adults. Nonetheless, juvenile offenders numbered in the tens of thousands across the state. In 1998, the state had fourteen hundred children locked up in training schools. With enactment of the 1998 Juvenile Justice Reform Act, North Carolina began to move away from the reformatory model and toward a new rehabilitation model. That year the Office of Juvenile Justice began to practice evidence-supported interventions. This nationally acclaimed comprehensive strategy encouraged community intervention as opposed to reformatory commitment.[11]

Increased emphasis on community-led intervention programs reduced the number of youths committed to juvenile detention centers by 77 percent. Between 2011–2013, the state closed three youth detention centers (YDCs), including Samarcand YDC, Swannanoa YDC, Camp Woodson, and Edgecombe YDC. Six YDC's (a total of 321 beds) remained in operation at a cost of more than $100,000 per bed per year. In the wake of these closures, the Justice Reinvestment Act of 2011 authorized a juvenile justice response, including the creation or expansion of multipurpose group homes, residential treatment providers, programs with parks and recreation offices, court-ordered parenting classes, mental health intervention, and crime prevention grants. By 2012, only three hundred of thirty thousand juvenile offenders remained in YDCs across the state.[12]

Budget cuts threatened the performance of the new reforms in juvenile justice. In 2013, the General Assembly downsized the cabinet-level Department of Juvenile Justice and Delinquency Prevention into the Division of Juvenile Justice and moved it into the Department of Public Safety. The downsizing ended special court services, cut prevention programs, and closed a correctional center. Youth advocates expressed concern that the downsizing would reverse a marked downward reduction in juvenile crime. Rob Thompson, executive director of the Covenant with North Carolina's Children, argued against the merger into Public Safety. "Preventing juvenile crime and rehabilitating youthful offenders demand a different model than adult corrections," Thompson argued, "we are concerned that the new structure will lead to a decreased focus on youth-centered programming." Linda W. Hayes, former secretary of the Department of Juvenile Justice and Delinquency Prevention, agreed that the merger was a move in the wrong direction. Hayes argued that the merger created problems in several major areas, including the fact that juvenile confidentiality prevented integration of data systems, managers were not trained to evaluate juvenile justice practices, and education programs differed, as adults focused on earning the GED while juveniles concentrated on finishing high school and attaining college educations. "North Carolina got it right" with the juvenile reforms of 1998–99, she argued, but the new merger proved a step backward. "Youths require different interventions applied in community-based settings," Hayes argued, "in concert with access to social, mental health, family and community help."[13]

Still, it remains to be seen if the state will authorize the resources that communities need to care for their high-risk youth. Rehabilitation, reinvestment, and realignment are cheaper than incarceration costs. But these strategies do not come for free. Rehabilitation requires funding for probation supervisors and community programs, costs that localities historically have found prohibitive. Realignment—that is, fiscal restructuring—requires creative financing in state and local budgets to pay for programs, and to provide proper expansion to county jails needed for short-term incarceration of serious youth offenders. Communities that lack financial resources will have no provision for youth offenders and no prevention programs for high-risk youth. State legislators should acknowledge the burden on localities and make state appropriations for rehabilitation. Otherwise, the state risks a return to nineteenth-century models of justice, when youth offenders landed in the adult justice system and high-risk youth simply wandered the streets. It was this reality that a century ago contributed to the white supremacist and eugenics policies that criminal justice authorities reject today.

NOTES

INTRODUCTION

1. "Margaret Pridgen," in Nell Battle Lewis, "Samarcand Arson Case, Carthage, N.C., May 18, 1931," in folder "Samarcand Arson Case," Nell Battle Lewis Papers, North Carolina State Archives, Raleigh, North Carolina (hereafter cited as Lewis Papers). A note on spelling: Scholars consistently reference the institution as "Samarcand Manor." For the sake of consistency, I have adopted that usage for this book. However, state records often reference the institution as "Samarkand Manor."

2. "Margaret Abernethy," "Marian Mercer," and "Virginia Hayes," in ibid.

3. John Wertheimer, Brian Luskey, et al. estimate the damages from local and national newspapers in "'Escape of the Match-Strikers': Disorderly North Carolina Women, the Legal System, and the Samarcand Arson Case of 1931," *North Carolina Historical Review* 75, no. 4 (October 1998): 444. For other article-length accounts of the trial, see Susan Cahn, "Spirited Youth or Fiends Incarnate: The Samarcand Arson Case and Female Adolescence in the American South," *Journal of Women's History* 9, no. 4 (winter 1998): 152–180; Annette Louise Bickford, "Imperial Modernity, National Identity and Capital Punishment in the Samarcand Arson Case, 1931," *Nations and Nationalism* 13, no. 3 (July 2007): 437–460.

4. For more elaborate detail on the case, see also Bickford, "Imperial Modernity, National Identity and Capital Punishment in the Samarcand Arson Case, 1931," 437–439, and Wertheimer, Luskey, et al., "Escape of the Match-Strikers," 435–436.

5. On Victorian-era reform, see Ruth M. Alexander, *The "Girl Problem": Female Sexual Delinquency in New York, 1900–1930* (Ithaca: Cornell Univ. Press, 1998); Mary E. Odem, *Delinquent Daughters: Protecting and Policing Adolescent Female Sexuality in the United States, 1885–1920* (Chapel Hill: Univ. of North Carolina Press, 1995); and Elizabeth Alice Clement, *Love for Sale: Courting, Treating, and Prostitution in New York City, 1900–1945* (Chapel Hill: Univ. of North Carolina Press, 2006). On southern women and reform, see Anne Firor Scott, *The Southern Lady: From Pedestal to Politics, 1830–1930* (1970; Charlottesville: Univ. Press of Virginia, 1995), 4–21; Laura F. Edwards, *Scarlett Doesn't Live Here Anymore: Southern Women in the Civil War Era* (Urbana: Univ. of Illinois Press, 2000), 5, 181. See also Glenda Elizabeth Gilmore, *Gender and Jim Crow: Women and the Politics of White Supremacy in North Carolina, 1896–1920* (Chapel Hill: Univ. of North Carolina Press, 1996); Jacquelyn Dowd Hall, *Revolt Against Chivalry: Jessie Daniel Ames and the Women's Campaign Against Lynching* (New York: Columbia Univ. Press, 1993); Anastatia Sims, *The Power of Femininity in the New South: Women's Organizations and Politics in North Carolina, 1880–1930*

(Columbia: Univ. of South Carolina Press, 1997); Marjorie Spruill Wheeler, *New Women of the New South: The Leaders of the Woman Suffrage Movement in the Southern States* (New York: Oxford Univ. Press, 1993); and Susan K. Cahn, *Sexual Reckonings: Southern Girls in a Troubling Age* (Boston: Harvard Univ. Press, 2007).

6. On the history of eugenics in the American South and in the United States, see Pippa Holloway, *Sexuality, Politics, and Social Control in Virginia, 1920–1945* (Chapel Hill: Univ. of North Carolina Press, 2006); Edward J. Larson, *Sex, Race, and Science: Eugenics in the Deep South* (Baltimore: Johns Hopkins Univ. Press, 1996), 40, 63; Anna Krome-Lukens, "The Reform Imagination: Gender, Eugenics, and the Welfare State in North Carolina, 1900–1940" (Ph.D. diss., University of North Carolina at Chapel Hill, 2014); Cahn, *Sexual Reckonings*, 156–180; Johanna Schoen, *Choice and Coercion: Birth Control, Sterilization, and Abortion in Public Health and Welfare* (Chapel Hill: Univ. of North Carolina Press, 2005), 75–138. See also Wendy Kline, *Building a Better Race: Gender, Sexuality, and Eugenics from the Turn of the Century to the Baby Boom* (Berkeley: Univ. of California Press, 2005); Katherine Castles, "Quiet Eugenics: Sterilization in North Carolina's Institutions for the Mentally Retarded, 1945–1965," *Journal of Southern History* 68, no. 4 (November 2002): 853; Edwin Black, *War Against the Weak: Eugenics and America's Campaign to Create a Master Race* (New York: Four Walls Eight Windows, 2003), xvi–xvii; Susan Currell and Christina Cogdell, eds., *Popular Eugenics: National Efficiency and American Mass Culture in the 1930s* (Athens: Ohio Univ. Press, 2006); Stefan Kuhl, *The Nazi Connection: Eugenics, American Racism, and German National Socialism* (New York: Oxford Univ. Press, 1994). Rickie Solinger provides an excellent history of the scholarship of reproductive politics in "Layering the Lenses: Toward Understanding Reproductive Politics in the United States," *Journal of Women's History* 25 (winter 2013): 101–112.

7. Cahn, *Sexual Reckonings*, 45; Alexander, *The "Girl Problem"*; Regina G. Kunzel, *Fallen Women, Problem Girls: Unmarried Mothers and the Professionalization of Social Work, 1890–1940* (New Haven: Yale Univ. Press, 1995).

8. Cahn, *Sexual Reckonings*, 44. On Nell Battle Lewis, see Alexander Leidholt, *Battling Nell: The Life of Southern Journalist Nell Battle Lewis* (Baton Rouge: Louisiana State Univ. Press, 2009); and Elizabeth Gillespie McRae, "To Save a Home: Nell Battle Lewis and the Rise of Southern Conservatism," *North Carolina Historical Review* 81, no. 3 (July 2004): 261–287. The Samarcand story has captured the imagination of historians. See Melton McLaurin and Anne Russell, *The Wayward Girls of Samarcand: A True Story of the American South* (Wilmington, N.C.: Bradley Creek Press, 2012); Cahn, "Spirited Youth or Fiends Incarnate"; Wertheimer, Luskey, et al., "Escape of the Match-Strikers"; and Bickford, "Imperial Modernity, National Identity and Capital Punishment in the Samarcand Arson Case, 1931."

9. Black, *War Against the Weak*, xvi–xvii; Currell and Cogdell, eds., *Popular Eugenics*; Kuhl, *The Nazi Connection*. See also Kline, *Building a Better Race*. On eugenics in the U.S. South, see Larson, *Sex, Race, and Science*, 1–4; Currell and Cogdell, eds., *Popular Eugenics*; Johanna Schoen, *Choice and Coercion*, 12–13; Katherine Castles, "Quiet Eugenics"; and Cahn, *Sexual Reckonings*, 158.

1. A PLACE FOR WHITE GIRLS

1. Quote is from "Home for Fallen Women," *Raleigh (N.C.) News and Observer*, March 4, 1917, 2; "Establishment of Home for Fallen Girls Advocated," *Raleigh (N.C.) News and Observer*, February

7, 1917, 2; Sallie Southall Cotten, *History of the North Carolina Federation of Women's Clubs, 1901–1925* (Raleigh, N.C.: Edwards and Broughton, 1925), 107–108; Kate Burr Johnson to Mrs. T. W. Lingle, May 1, 1917, G. Hope Summerell Chamberlain Papers, box 3, David M. Rubenstein Rare Book and Manuscript Collection, Duke University, Durham, North Carolina (hereafter referred to as "Rubenstein Collection"); R.D.W. Connor, ed., *North Carolina Manual* (Raleigh, N.C.: Edwards and Broughton, 1917), 18. "To Establish a State Home and Industrial School For Girls and Women," SB 729, HB 1918, box 5, Records of the North Carolina General Assembly, Session 1917, North Carolina State Archives, Raleigh, North Carolina.

2. Kunzel, *Fallen Women, Problem Girls,* 2–3.

3. Walter Rauschenbusch, *Christianity and the Social Crisis,* Library of Theological Ethics (Louisville, Ky.: Westminster John Knox Press, 1992), 216–218; Reverend A. A. McGeachy, "The Face of Jesus: A Sermon Preached by Rev. A. A. McGeachy, D.D., on March 15th, in the Second Church, Charlotte, N.C.," reprint from the *Presbyterian Standard,* March 25, 1914, in G. Hope Summerell Chamberlain Papers, Writings: Miscellaneous, box 18.

4. McGeachy, "The Face of Jesus," 3, 8.

5. Ibid., 6.

6. Kunzel, *Fallen Women, Problem Girls,* 20–22; Judith R. Walkowitz, *City of Dreadful Delight: Narratives of Sexual Danger in Late-Victorian London* (Chicago: Univ. of Chicago Press, 1992), 20; Susanna Rowson, *Charlotte Temple,* ed. Cathy N. Davidson (Oxford: Oxford Univ. Press, 1986); Stephen Crane, *Maggie: A Girl of the Streets, and Other New York Writings* (1893; reprint, New York: Modern Library, 2001).

7. McGeachy, "The Face of Jesus," 6–7.

8. Ibid., 7.

9. Lori D. Ginzberg, *Women and the Work of Benevolence: Morality, Politics, and Class in the Nineteenth-Century United States* (New Haven: Yale Univ. Press, 1992), 34–37; see also Gilmore, *Gender and Jim Crow,* introduction.

10. Cotten, *History of the North Carolina Federation of Women's Clubs,* 2; Sims, *The Power of Femininity in the New South,* 3–4, 167.

11. Hope Summerell Chamberlain, *This Was Home* (Chapel Hill: Univ. of North Carolina Press 1938), 126, 137, 139.

12. Sims, *The Power of Femininity,* 2, quote on pages 35, 156; Hall, *Revolt Against Chivalry,* 151–152; Jacquelyn Dowd Hall, "'The Mind That Burns in Each Body': Women, Rape, and Racial Violence," in *Powers of Desire: The Politics of Sexuality,* ed. Ann Snitow, Christine Stansell, and Sharon Thompson (New York: Monthly Review, 1983), 334.

13. "Suffrage: Abt. 10 short years ago," [circa 1929], in "Writings, Miscellaneous," G. Hope Sumerell Chamberlain Papers, 5.

14. Ibid., 7; Chamberlain, *This Was Home,* 32–34, 297.

15. Cotten, *History of the North Carolina Federation of Women's Clubs,* 107–110; See also Kate Burr Johnson to Mrs. T. W. Lingle, May 1, 1917, Correspondence, 1915–1920, box 3, G. Hope Summerell Chamberlain Papers.

16. For biographical information on Kate Burr Johnson, see "Mrs. Kate Burr Johnson," "Mrs. Clarence Johnson's Administration as President of the North Carolina Federation of Women's Clubs, 1917–1919," and "Elected Director Division of Child Welfare, State Board of Charities and Public Welfare, June 1919," all in "Speeches, Reports and Miscellaneous," in Kate Ancrum Burr

Johnson Papers, 91.1.c, Joyner Library Special Collections, East Carolina University, Greenville, North Carolina (hereafter referred to as the Kate Ancrum Burr Johnson Papers). Quote from "Mrs. Clarence Johnson's Administration as President of the North Carolina Federation of Women's Clubs, 1917–1919," 2. Mrs. Thomas W. Lingle to Hope Summerell Chamberlain and Adelaide L. Fries to Hope Summerell Chamberlain, both dated April 19, 1917, Correspondence, 1915–1920, box 3, G. Hope Summerell Chamberlain Papers; Nancy F. Cott, *The Grounding of Modern Feminism* (New Haven: Yale Univ. Press, 1987), 216–218.

17. Kate Burr Johnson, "Problems of Delinquency Among Girls," *Journal of Social Hygiene* 12, no. 7 (October 1926): 385–386, 396–397.

18. Anna Krome-Lukens, "A Great Blessing to Defective Humanity: Women and the Eugenics Movement in North Carolina, 1910–1940" (Master's thesis, University of North Carolina at Chapel Hill, 2009), 32–37.

19. Allan M. Brandt, *No Magic Bullet: A Social History of Venereal Disease in the United States Since 1880* (New York: Oxford Univ. Press, 1987), 87; Martha P. Falconer, "Causes of Delinquency Among Girls," *Annals of the American Academy of Political and Social Science* 36, no. 1 (July 1910): 77–79. See also Martha P. Falconer, "Report of the Committee on Social Hygiene of the National Conference of Charities and Correction," *Social Hygiene* 1 (September 1915): 520–521; Martha P. Falconer, "Industrial Schools for Girls and Women," *Social Hygiene* 3, no. 3 (July 1917): 323–330; Martha P. Falconer, "The Part of the Reformatory Institution in the Elimination of Prostitution," *Social Hygiene* 5, no. 1 (January 1919): 3–6. On social hygiene and social reform, see Kristin Luker, "Sex, Social Hygiene, and the State: The Double-Edged Sword of Social Reform," *Theory and Society* 27, no. 5 (October 1998): 601–634.

20. Brandt, *No Magic Bullet*, 165; American Social Hygiene Association, "The American Social Hygiene Association, Inc. [1918]," *Journal of Social Hygiene* 4 (1918): 138a; Martha P. Falconer, "The Segregation of Delinquent Women and Girls as a War Problem," in *The Annals: War Relief Work*, ed. J. P. Lichtenberger (Philadelphia: American Academy of Political and Social Science, 1918): quotes on pages 160–162.

21. Currell and Cogdell, eds., *Popular Eugenics*, 125–126, 165.

22. Falconer, "The Segregation of Delinquent Women and Girls as a War Problem," 166.

23. Kunzel, *Fallen Women, Problem Girls*, 51–53.

24. "Welfare Measure Is Introduced in Senate by Scales," *Raleigh (N.C.) News and Observer*, January 16, 1917, 6; Cotten, *History of the North Carolina Federation of Women's Clubs*, 101. "An Act to Repeal Chapter 85 of the Revisal of 1905 and Substitute in lieu thereof a New Chapter Creating 'The State Board of Charities and Public Welfare' and Defining its Duties and Powers," SB 543, HB 1018, box 11, Records of the North Carolina General Assembly, Session Records 1917, North Carolina State Archives, Raleigh, North Carolina; T. W. Bickett to Hope Summerell Chamberlain, May 16, 1918, and Reverend A. A. McGeachy to Hope Summerell Chamberlain, December 16, 1920, both in Correspondence, 1915–1920, box 3, G. Hope Summerell Papers.

25. Thomas Walter Bickett to Hope Summerell Chamberlain, May 16, 1918, Correspondence, 1951–1918, box 3, G. Hope Summerell Chamberlain Papers; "Will Hold Hearing for Woman's Home at 3:30 pm Today," *Raleigh (N.C.) News and Observer*, February 6, 1917, 3. For bills, see "To Establish a State Home and Industrial School For Girls and Women," SB 729, HB 1918, in Senate Bills, 562–755, box 5, and "To Provide for the issue of Twenty-five Thousand Dollars of bonds for the Purpose of Purchasing Land and Erecting Buildings for a State Home and Training School for

Girls and Women," SB 1541, HB 1856, box 16, Records of the North Carolina General Assembly, Session Records 1917, North Carolina State Archives, Raleigh, North Carolina.

26. Young Women's Christian Association of Charlotte, January 27, 1917, Board of Aldermen of City of Goldsboro, February 1, 1917, Ministerial Union of Elizabeth City, February 2, 1917, St. Peter's Episcopal Church, Washington, N.C., February 2, 1917, W.C.T.U., Spring Hope, North Carolina, n.d., Executive Committee of Western North Carolina Women's Missionary Society of the M. E. Episcopal Church, South at Newton, N.C., February 10, 1917, W.C.T.U., North Carolina, February 1, 1917, Pasquotank County and City of Elizabeth City Citizens at a Community Service Day, February 3, 1917, telegrams to C. G. Wright from citizens in Charlotte, February 5, 1917, all in "Petitions (Schools—State Home & Training School for Girls and Women)," box 19, North Carolina General Assembly Session Records, 1917; "Home for Women Is Endorsed by Women," *Raleigh (N.C.) News and Observer,* January 26, 1917, 1, col. 4.

27. "Bill to Create Home Is Passed," *Charlotte (N.C.) News,* March 7, 1917, 2, col. 3; "Establishment of Home for Fallen Girls Advocated," *Raleigh (N.C.) News and Observer,* February 7, 1917, 2.

28. "Establishment of Home for Fallen Girls Advocated," *Raleigh (N.C.) News and Observer,* February 7, 1917, 2.

29. Ibid.

30. Ibid.

31. "The Wayward Child," *Charlotte (N.C.) Sunday Observer,* February 25, 1917, 27, cols. 1–2; see also "Women's Home Being Urged," *Charlotte (N.C.) News,* March 5, 1917, 11, col. 3.

32. "Pegram Bill for City Government Now Among Laws," *Raleigh (N.C.) News and Observer,* March 4, 1917, 1; "Home for Fallen Women," *Raleigh (N.C.) News and Observer,* March 4, 1917, 2; S. Lillian Clatton [*sic,* Clayton], "The Social Training of the Private Duty Nurse," *American Journal of Nursing* 17, no. 6 (March 1917): 506–508, quote on page 508; "Birdie Dunn," obituaries, *American Journal of Nursing* 46, no. 9 (September 1946): 640.

33. "Pegram Bill for City Government Now Among Laws," *Raleigh (N.C.) News and Observer,* March 4, 1917, 1; "Home for Fallen Women," *Raleigh (N.C.) News and Observer,* March 4, 1917, 2

34. William A. Link, *The Paradox of Southern Progressivism, 1880–1930* (Chapel Hill: Univ. of North Carolina Press, 1992), 62–63, quote on page 294; William A. Link, "Pre-bureaucratic Social Policy and the Emergence of Reform," in *The Wilson Era: Essays in Honor of Arthur S. Link,* ed. John Milton Cooper Jr. and Charles E. Neu (Arlington Heights, Ill.: Harlan Davidson, 1991).

35. "Reformatory for Women," *Raleigh (N.C.) News and Observer,* March 6, 1917, 10, col. 4. Murphy's account of the clock is in "Bill to Create Home Is Passed," *Charlotte (N.C.) News,* March 7, 1917, 2, col. 3. "An Act to Repeal Chapter 85 of the Revisal of 1905 and Substitute in Lieu Thereof a New Chapter Creating 'The State Board of Charities and Public Welfare' and Defining its Duties & Powers," SB 543, HB 1018, box 11; "An Act to Establish a State Home and Industrial School for Girls and Women," SB 729, HB 1918, box 5; "An Act to Provide for the issue of Twenty-five Thousand Dollars of Bonds for the Purpose of Purchasing Land and Erecting Buildings for a State Home and Training School for Girls and Women," SB 1541, HB 1856, box 16, all in Records of the North Carolina General Assembly, Session Records 1917, North Carolina State Archives, Raleigh, North Carolina. The General Assembly also passed an "Act to Establish Reformatories or Homes for Fallen Women" that granted counties and cities the "right and power" to establish local reformatories in cities with populations exceeding twenty thousand people, and to tax for that purpose, SB 619, HB 1855, box 12, Records of the North Carolina General Assembly, Ses-

sion Records 1917, North Carolina State Archives, Raleigh, North Carolina. See also "Samarcand Opens Door of Hope to 1000th Girl in Tenth Year," *Raleigh (N.C.) News and Observer,* October 7, 1928, 9, cols 1–8.

2. IN DEFENSE OF THE NATION

1. On the "girl problem," see Alexander, *The "Girl Problem"*; Clement, *Love for Sale*; Nancy K. Bristow, *Making Men Moral: Social Engineering During the Great War* (New York: New York Univ. Press, 1996); and Cahn, *Sexual Reckonings,* 5, 10,14, 23, 45.

2. Brandt, *No Magic Bullet,* 7–9, 14–18, 30–31. On sexuality, adolescence and venereal disease in North Carolina, see also Cahn, *Sexual Reckonings,* 182–188.

3. Brandt, *No Magic Bullet,* 7–9, 14–18, 30–31.

4. Bristow, *Making Men Moral,* xviii, 3, 8–11, 19, 23, 92–95.

5. Ibid., 99–104, 107–109, 114.

6. Brandt, *No Magic Bullet,* 8, 57.

7. *Manual for the Various Agents of the United States Interdepartmental Social Hygiene Board* (Washington, D.C.: GPO, 1920), 16–19.

8. James A. Keiger, "Introductory or General Statement in Behalf of State Legislation for the Control of Social Diseases," 1–2 (quote on page 1), box 13, Records of the State Board of Health, 1919, North Carolina State Archives, Raleigh, North Carolina.

9. James A. Keiger, "Introductory or General Statement in Behalf of State Legislation for the Control of Social Diseases," 1–2 (quote on page 1), box 13, Records of the State Board of Health, 1919.

10. Ibid., 3; on incidences of syphilis by sex, see "Dr. James A. Keiger to Dr. W. S. Rankin," January 23, 1919, box 13, Records of the State Board of Health, 1919.

11. "Dr. Carl D. Reynolds to Dr. W. S. Rankin," December 29, 1917, box 12, Records of the State Board of Health, 1918.

12. Link, *The Paradox of Southern Progressivism,* 6–7, and see footnotes on pages 287–288 and 387.

13. Sections 1–9, "An Ordinance for the Control of Venereal Disease," March 1918, Charlotte, North Carolina, box 12, Records of the State Board of Health, 1918; see also James A. Keiger, "Health Legislation by the General Assembly of 1919," 4–5, box 13, Records of the State Board of Health, 1919.

14. Sections 1–9, "An Ordinance for the Control of Venereal Disease," March 1918.

15. Brandt, *No Magic Bullet,* 40–43; "Chapter 62: An Act to Amend the Health Laws of North Carolina," *Public Laws and Resolutions of the State of North Carolina, passed by the General Assembly at its Session of 1911* (Raleigh, N.C.: Uzzell, 1911), 203–218, see 203–204 (volumes in this series hereafter referred to as *Public Laws and Resolutions of the State of North Carolina*); "Chapter 181: An Act to Amend the Health Laws of North Carolina," *Public Laws and Resolutions of the State of North Carolina* (1913), 277–281, see 277 and 279.

16. "Influenza" broadside, ordinance passed by the community Board of Directors of Oakdale Community, Alamance County, North Carolina, October 18, 1918; "Report of the Columbus County Board of Health," November 1, 1918, Whiteville, North Carolina; Ordinance of Lexington,

North Carolina, October 3, 1918; E. F. Long to Dr. W. S. Rankin, Secretary, State Board of Health, October 3, 1918, all in box 12, Records of the State Board of Health, 1918.

17. In 1911 the General Assembly endowed the State Board of Health with the power to regulate "where there is no regularly organized local board of health." See "Chapter 62: An Act to Amend the Health Laws of North Carolina," *Public Laws and Resolutions of the State of North Carolina* (1911), 203–204. See also "Chapter 181: An Act to Amend the Health Laws of North Carolina," *Public Laws and Resolutions of the State of North Carolina* (1913), 277. In 1917 the General Assembly allowed the State Board of Health to pass "minimum regulations" that local communities might adopt "as in the judgment of the authorities of such counties, towns and cities seem necessary." See "Chapter 263: An Act to Prevent and Control the Occurrence of Certain Infectious Diseases in North Carolina," *Public Laws and Resolutions of the State of North Carolina* (1917), 532–536, quote on 535.

18. Brandt, *No Magic Bullet,* 59–70, quote on page 73.

19. James A. Keiger, "Health Legislation by the General Assembly of 1919," 4, box 13, Records of the State Board of Health, 1919.

20. "War on Venereal Disease to Continue," *Raleigh (N.C.) News and Observer,* February 2, 1919, 11.

21. Keiger, "Health Legislation by the General Assembly of 1919," 4–5. The North Carolina General Assembly passed these acts in a series of sections contained in two chapters, "Chapter 214: An Act to Obtain Reports of Persons Infected with Venereal Disease" and "Chapter 215: An Act for the Repression of Prostitution," in *Public Laws and Resolutions of the State of North Carolina* (1919), 418–421. In 1921 the North Carolina General Assembly amended the laws of punishment to specify that any person arrested for prostitution or the aiding and abetting prostitution would be charged with a misdemeanor. See "An Act to Amend Chapter 215, Public Laws of 1919, Relating to the Repression of Prostitution," *Public Laws and Resolutions of the State of North Carolina* (1921), 359–360.

22. Kathy Peiss, *Cheap Amusements: Working Women and Leisure in Turn-of-the-Century New York* (Philadelphia: Temple Univ. Press, 1986), 110; Bristow, *Making Men Moral,* 117; Clement, *Love for Sale,* 5–6, 14–15, 48–49; and Alexander, *The "Girl Problem,"* 3, 12–14, 59, 62–63; Cahn, *Sexual Reckonings,* 182–188.

23. Millard Knowlton to Dr. T. F. Wickliffe, October 16, 1919, and T. F. Wickliffe to Dr. B. E. Washburn, October 17, 1919, both in box 13, Records of the State Board of Health, 1919.

24. "Field Agent's Report on the Repression of Prostitution, May 1st to October 1st, 1919," 4, box 13, Records of the State Board of Health, 1919.

25. Ibid., 1–3, 5.

26. State Health Officer to T. F. Wickliffe, October 21, 1919 and "Tentative Suggestions for Rules and Regulations for the Control of Venereal Diseases," 1–4, both in box 13, Records of the State Board of Health, 1919.

27. Venereal Disease Advisory Committee, "Principles in Syphilis Therapy, Technic of Treatment, and Sterilization," April 30, 1931, 1–5, box 38, Records of the State Board of Health, 1932–1935.

28. Ibid., 5–6.

29. W. S. Blakeney to R. F. Beasley, February 2, 1920; "United States Interdepartmental Social Hygiene Board Request for Maintenance," January 1, 1920; and T. A. Storey to Agnes B. Mac-

Naughton, March 3, 1920, all in "Samarcand Manor, 1918–1924," box 164, State Board of Charities and Public Welfare, Institutions and Corrections, State Charitable, Penal and Correctional Institutions, Samarcand, State Archives of North Carolina, Raleigh, North Carolina.

30. "United States Interdepartmental Social Hygiene Board Request for Maintenance," January 1, 1920; "Field Agent's Report on the Repression of Prostitution, May 1st to October 31st, 1919," 2, 5–8; Commissioner to Mrs. J. R. Chamberlain, March 11, 1920; and Clerk of the Wake County Superior Court to Samarcand Manor, January 13, 1920, all in "Samarcand Manor, 1918–1924," box 164, State Board of Charities and Public Welfare, Institutions and Corrections, State Charitable, Penal and Correctional Institutions, Samarcand, State Archives of North Carolina, Raleigh, North Carolina.

31. Cahn, *Sexual Reckonings,* 92–93; Tanya Smith Brice, "The Treatment of African American Girls in Progressive Era North Carolina's Juvenile Justice System, 1890–1930" (Ph.D. diss., School of Social Work, University of North Carolina at Chapel Hill, 2003), 8–14, 28–29, 66–70; North Carolina State Board of Charities and Public Welfare, *Biennial Report of the North Carolina State Board of Charities and Public Welfare, 1926–1928* (Raleigh, N.C.: Mitchell Printing Co.), 111.

3. HOW TO MAKE BAD GIRLS GOOD

1. Agnes MacNaughton's name frequently appears (incorrectly) as "McNaughton." Agnes MacNaughton, "A Glimpse of Life at Samarcand Manor," circa 1920, Samarcand Manor Scrapbook Collection, Records of Samarcand Manor, Division of Adult Correction and Juvenile Justice, Department of Public Safety, Samarcand Manor School, Eagle Springs, North Carolina, 1–2 (hereafter referred to as Samarcand Manor Scrapbook Collection). See also Agnes MacNaughton to Mrs. Clarence Johnson, September 23, 1920, in "Samarcand Manor, 1918–1924," box 164, State Board of Charities and Public Welfare, Institutions and Corrections, State Charitable, Penal and Correctional Institutions, Samarcand, State Archives of North Carolina, Raleigh, North Carolina.

2. Fred A. Olds, "Samarcand Manor: An Institution that Rescues Delinquent Girls from the Gutter and Cleanses Them," *Orphans' Friend and Masonic Journal,* December 17, 1920, 1 col. 1

3. Sims, *The Power of Femininity,* 2, 35, 156; Hall, *Revolt Against Chivalry,* 151–152; Hall, "'The Mind That Burns in Each Body,'" in Snitow, Stansell, and Thompson, eds., *Powers of Desire,* 334.

4. Agnes MacNaughton, "Samarcand Manor and Its Work," 1920, black scrapbook in Samarcand Manor Scrapbook Collection, Eagle Springs, North Carolina.

5. Agnes MacNaughton to Mrs. Clarence Johnson, September 27, 1921, in "Samarcand Manor, 1918–1924," box 164, State Board of Charities and Public Welfare, Institutions and Corrections, State Charitable, Penal and Correctional Institutions, Samarcand, State Archives of North Carolina, Raleigh, North Carolina.

6. W. S. Blakeney to R. F. Beasley, February 2, 1920, in "Samarcand Manor, 1918–1924," box 164, State Board of Charities and Public Welfare, Institutions and Corrections, State Charitable, Penal and Correctional Institutions, Samarcand, State Archives of North Carolina, Raleigh, North Carolina.

7. Agnes MacNaughton to R. F. Beasley, February 16, 1920, in "Samarcand Manor, 1918–1924," box 164, State Board of Charities and Public Welfare, Institutions and Corrections, State Charitable, Penal and Correctional Institutions, Samarcand, State Archives of North Carolina, Raleigh, North Carolina.

8. Beasley to W. S. Blakeney, February 28, 1920, in "Samarcand Manor, 1918–1924," box 164, State Board of Charities and Public Welfare, Institutions and Corrections, State Charitable, Penal and Correctional Institutions, Samarcand, State Archives of North Carolina, Raleigh, North Carolina.

9. T. A. Storey to Agnes B. MacNaughton, March 3, 1920; R. F. Beasley to L. A. Storey, February 28, 1920; and "Report for Biennial Period Ending November 1920," all in "Samarcand Manor, 1918–1924," box 164, State Board of Charities and Public Welfare, Institutions and Corrections, State Charitable, Penal and Correctional Institutions, Samarcand, State Archives of North Carolina, Raleigh, North Carolina.

10. Agnes MacNaughton to R. F. Beasley, February 16, 1920, and Agnes B. MacNaughton to Mrs. Clarence Johnson, September 23, 1920, both in "Samarcand Manor, 1918–1924," box 164, State Board of Charities and Public Welfare, Institutions and Corrections, State Charitable, Penal and Correctional Institutions, Samarcand, State Archives of North Carolina, Raleigh, North Carolina.

11. T. E. Browne, "Visit to Samarcand," *Raleigh (N.C.) News and Observer,* April 26, 1920.

12. Ibid.

13. Ibid.

14. Ibid.

15. Commissioner Johnson responded by saying that girls at Samarcand were fortunate for the second chance. She argued that other states placed girls in jails or other penal institutions, while North Carolina provided a school in which they were subjected to discipline and training that they failed to get in their homes. "My advice to you" Johnson recommended to this mother, "is to be very thankful that as long as your daughter has gotten into trouble that she could be sent to such as place as Samarcand." Parkton, N.C. [name redacted] to Mrs. Kate Burr Johnson, March 28, 1929, and Kate Burr Johnson to Parkton, N.C. [name redacted], March 26, 1929, both in "Investigations and Complaints, etc. 1926–1934," Institutions and Corrections, State Charitable, Penal and Correctional Institutions, Records of the State Board of Charities and Public Welfare, series 97, box 165, North Carolina State Archives, Raleigh, North Carolina.

16. Fred A. Olds to Agnes MacNaughton, December 12, 1919, miscellaneous scrapbook, Samarcand Manor Scrapbook Collection, Eagle Springs, North Carolina; Fred A. Olds, "Samarcand Manor: An Institution that Rescues Delinquent Girls from the Gutter and Cleanses Them," December 17, 1920, *Raleigh (N.C.) News and Observer,* 1, col. 1.

17. Fred A. Olds, "Samarcand Manor: An Institution that Rescues Delinquent Girls from the Gutter and Cleanses Them," December 17, 1920, *Raleigh (N.C.) News and Observer,* 1 col. 1.

18. Ibid.

19. Agnes MacNaughton, "A Glimpse of Life at Samarcand Manor," 1920, and "Samarcand Manor," circa 1922, miscellaneous scrapbook, Samarcand Manor Scrapbook Collection, Eagle Springs, North Carolina. The Goldsboro speech quote appears on page 5.

20. MacNaughton, "A Glimpse of Life at Samarcand Manor,'" scrapbook, 4, Samarcand Manor Scrapbook Collection, Eagle Springs, North Carolina; Fred A. Olds, "Samarcand Manor: An Institution that Rescues Delinquent Girls from the Gutter and Cleanses Them." December 17, 1920, *Raleigh (N.C.) News and Observer,* 1, col. 1.

21. W. A. Blair to Mrs. Clara Johnson, May 24, 1921, and Kate Burr Johnson to Mrs. Mary O. Linton, November 25, 1921, both in "Samarcand Manor, 1918–1924," box 164, State Board of Charities and Public Welfare, Institutions and Corrections, State Charitable, Penal and Correctional Institutions, Samarcand, State Archives of North Carolina, Raleigh, North Carolina.

22. Report of Inspection of Samarcand Manor, July 4–7, 1921, in "Samarcand Manor, 1918–1924," box 164, State Board of Charities and Public Welfare, Institutions and Corrections, State Charitable, Penal and Correctional Institutions, Samarcand, State Archives of North Carolina, Raleigh, North Carolina.

23. Grace A. Reeder, Report of Inspection of Samarcand Manor, July 4–7, 1921, in "Samarcand Manor, 1918–1924," box 164, State Board of Charities and Public Welfare, Institutions and Corrections, State Charitable, Penal and Correctional Institutions, Samarcand, State Archives of North Carolina, Raleigh, North Carolina.

24. Ibid.

25. Mrs. C. A. Loop, "A Home and Not a Prison: A School Where the Honor System Rules," circa 1922, newspaper clipping from miscellaneous scrapbook, Samarcand Manor Scrapbook Collection, Eagle Springs, North Carolina.

26. Ibid.; Agnes MacNaughton, "A Glimpse of Life at Samarcand Manor," circa 1920, Samarcand Manor Scrapbook Collection, Eagle Springs, North Carolina, 3.

27. "Student Government Don'ts," 1922, miscellaneous scrapbook, Samarcand Manor Scrapbook Collection, Eagle Springs, North Carolina.

28. Ibid., 4; "Student Government," 1922, miscellaneous scrapbook, Samarcand Manor Scrapbook Collection, Eagle Springs, North Carolina.

29. "Samarcand Manor: Re: May McCubbins, June 29, 1926," in "Samarcand Manor: Investigations and Complaints, etc. 1926–1934," Institutions and Corrections, State Charitable, Penal, and Correctional Institutions, Records of the State Board of Public Welfare, series 97, box 165, North Carolina State Archives, Raleigh, North Carolina.

30. Ibid. The parent's name is redacted for the protection of the child, but correspondence is headed "Parkton Mattress Co. of Parkton, N.C." [Unnamed parent] to Mrs. Kate Burr Johnson, circa March 1929, in "Samarcand Manor: Investigations and Complaints, etc. 1926–1934," Institutions and Corrections, State Charitable, Penal and Correctional Institutions, Records of the State Board of Public Welfare, series 97, box 165, North Carolina State Archives, Raleigh, North Carolina.

31. "Annual Field Day," 1922, miscellaneous scrapbook, Samarcand Manor Scrapbook Collection, Eagle Springs, North Carolina.

32. Paula D. Welch, *History of American Physical Education and Sport*, 3rd ed. (Springfield, Ill.: Thomas, 2004), 84.

33. "Annual Field Day," 1922, miscellaneous scrapbook, Samarcand Manor Scrapbook Collection, Eagle Springs, North Carolina; Welch, *History of American Physical Education and Sport*, 84; Paul Atkinson, "The Feminist Physique: Physical Education and the Medicalization of Women's Education," in *From "Fair Sex" to Feminism: Sport and the Socialization of Women in the Industrial and Post-Industrial Eras*, edited by J. A. Mangan and Roberta J. Park (London: Cass, 1987), 38–57. Vital energy is discussed on page 42.

34. "The Fire," miscellaneous scrapbook, 1922, Samarcand Manor Scrapbook Collection, Eagle Springs, North Carolina.

4. SUDDENLY PROCLAIMED UNFIT

1. "Kate Burr Johnson: She's a Tradition in Her Own Lifetime," circa 1964, and "University Sorority Honors Kate Burr Johnson," *Raleigh (N.C.) News and Observer,* April 8, 1954, both in "News-

paper Clippings," 91.1.b, Kate Ancrum Burr Johnson Papers. See also "Mrs. Kate Burr Johnson," biographical sketch, "Speeches, Reports and Miscellaneous," 91.1.c, Kate Ancrum Burr Johnson Papers. On North Carolina women reformers, see Krome-Lukens, "The Reform Imagination," and Krome-Lukens, "A Great Blessing to Defective Humanity," 5–9, 32–40.

2. "Mrs. Johnson's Citation," n.d., n.p., in "Newspaper Clippings," 91.1.b, Kate Ancrum Burr Johnson Papers; Krome-Lukens, "A Great Blessing to Defective Humanity," 32–40.

3. Alexander, *The "Girl Problem,"* 37–39, 44.

4. North Carolina State Board of Charities and Public Welfare, *Biennial Report of the North Carolina State Board of Charities and Public Welfare, 1922–1924* (Raleigh, N.C.: The Board, 1924), 9.

5. Alexander, *The "Girl Problem,"* 59–61; Krome-Lukens, "A Great Blessing to Defective Humanity," 3–7.

6. North Carolina State Board of Charities and Public Welfare, *Biennial Report of the North Carolina State Board of Charities and Public Welfare, 1920–1922* (Raleigh, N.C.: The Board, 1922), 7–12, 40a, 46.

7. Larson, *Sex, Race, and Science,* 3; Daniel J. Kevles, *In the Name of Eugenics: Genetics and the Uses of Human Heredity* (Cambridge: Harvard Univ. Press, 1998), 14; North Carolina State Board of Charities and Public Welfare, *Biennial Report, 1922–1924,* 11, 134.

8. William Fielding Ogburn, Review of *Applied Eugenics* by Paul Popenoe and Roswell Hill Johnson, *Political Science Quarterly* (September 1921), 533–534; William Fielding Ogburn, *Social Change with Respect to Culture and Original Nature* (New York: Viking, 1928), 2, 11–12, 20–22, 20–251; Currell and Cogdell, eds., *Popular Eugenics,* 5.

9. Johnson, "Problems of Delinquency Among Girls," 385.

10. Brice, "The Treatment of African American Girls in Progressive Era North Carolina's Juvenile Justice System," 8–14, mission statement on pages 13–14.

11. Ibid., 8–14, 68–70; North Carolina State Board of Charities and Public Welfare, *Biennial Report, 1926–1928,* 111.

12. Brice, "The Treatment of African American Girls in Progressive Era North Carolina's Juvenile Justice System," 28–29, 66.

13. North Carolina State Board of Charities and Public Welfare, *Biennial Report, 1926–1928,* 98, quote on pages 101, 108. See also Yolanda Burwell, "Lawrence A. Oxley: Defining State Public Welfare among African Americans," in *African American Leadership: An Empowerment Tradition in Social Welfare History,* ed. I. B. Carlton-LaNey (Washington, DC: NASW Press, 2001), 99–110.

14. Winston-Salem State University was originally named Winston-Salem Teachers College. "Kate Burr Johnson Praised by Negroes," n.d., n.p., "Newspaper Clippings," 91.1.b, Kate Ancrum Burr Johnson Papers.

15. Kate Burr Johnson, "Problems of Delinquency Among Girls," 386, 389–391.

16. Ibid., 389–391.

17. North Carolina State Board of Charities and Public Welfare, *Biennial Report, 1922–1924,* 10.

18. Odem, *Delinquent Daughters,* 169; North Carolina State Board of Charities and Public Welfare, *Biennial Report, 1922–1924,* 19–21.

19. Odem, *Delinquent Daughters,* 169; North Carolina State Board of Charities and Public Welfare, *Biennial Report, 1922–1924,* 19–21; "Chapter 260: An Act to Aid Needy Orphan Children in the Homes of Worthy Mothers," *Public Laws and Resolutions of the State of North Carolina* (1923), 583–584.

20. North Carolina State Board of Charities and Public Welfare, *Biennial Report, 1922–1924*, 30–36, quote on page 36.

21. North Carolina State Board of Charities and Public Welfare, *Biennial Report, 1926–1928*, 30, quote on pages 34, 35.

22. "Proceedings of the Conference Held by the Committee on the Care and Training of Delinquent Women and Girls of the National Committee on Prisons and Prison Labor," in "Samarcand Manor, 1918–1924," box 164, State Board of Charities and Public Welfare, Institutions and Corrections, State Charitable, Penal and Correctional Institutions, Samarcand, State Archives of North Carolina, Raleigh, North Carolina.

23. See "Daily Surface Data" for Pinehurst and Southern Pines Stations, Moore County, January 1922, at the National Climatic Data Center, Climatic Data Online, www.ncdc.noaa.gov/cdo -web/, last accessed November 1, 2010.

24. "Proceedings of the Conference Held by the Committee on the Care and Training of Delinquent Women and Girls of the National Committee on Prisons and Prison Labor," in "Samarcand Manor, 1918–1924," box 164, State Board of Charities and Public Welfare, Institutions and Corrections, State Charitable, Penal and Correctional Institutions, Samarcand, State Archives of North Carolina, Raleigh, North Carolina.

25. Ibid.

26. "Proceedings of the Conference Held by the Committee on the Care and Training of Delinquent Women and Girls of the National Committee on Prisons and Prison Labor," in "Samarcand Manor, 1918–1924," box 164, State Board of Charities and Public Welfare, Institutions and Corrections, State Charitable, Penal and Correctional Institutions, Samarcand, State Archives of North Carolina, Raleigh, North Carolina, pages 7–8, quote on page 8.

27. Ibid., 7–8.

28. Alexander, *The "Girl Problem,"* 89; North Carolina State Board of Charities and Public Welfare, *Biennial Report, 1922–1924*, quote on page 11. On eugenics and the U.S. South, see Larson, *Sex, Race, and Science,* and Steven Noll, *Feeble-Minded in Our Midst: Institutions for the Mentally Retarded in the South, 1900–1940* (Chapel Hill: Univ. of North Carolina Press, 1995). Also, on eugenics and mental hygiene, see Kevles, *In the Name of Eugenics*; Kline, *Building a Better Race*; Nancy Ordover, *American Eugenics: Race, Queer Anatomy, and the Science of Nationalism* (Minneapolis: Univ. of Minnesota Press, 2003); and Alexandra Stern, *Eugenic Nation: Faults and Frontiers of Better Breeding in Modern America* (Berkeley: Univ. of California Press, 2005).

29. Kate Burr Johnson, "Mental Deficiency Speech to Medical Society," Kate Ancrum Burr Johnson Papers, 15–16.

30. Ibid.; North Carolina State Board of Charities and Public Welfare, *Biennial Report, 1922–1924*, 45, quote on page 128.

31. Johnson estimated that fifty-five thousand North Carolinians were of "lower levels of intelligence." See Kate Burr Johnson, "State Welfare Program as It Relates to Mental Diseases and Crime," speech to the Medical Society, Lenoir, North Carolina, September 1927, "Speeches, Reports and Miscellaneous," 91.1.c, Kate Ancrum Burr Johnson Papers, 14; North Carolina State Board of Charities and Public Welfare, *Biennial Report, 1922–1924*, 132.

32. North Carolina State Board of Charities and Public Welfare, *Biennial Report, 1922–1924*, 144–145.

33. Kevles, *In the Name of Eugenics*, 79–81; North Carolina State Board of Charities and Public Welfare, *Biennial Report, 1922–1924*, 144–145.

34. Johnson, "State Welfare Program as It Relates to Mental Diseases and Crime," 27–28; North Carolina State Board of Charities and Public Welfare, *Biennial Report, 1920–1922*, 39–40.

35. Ibid., 25–26.

36. Johnson, "State Welfare Program as It Relates to Mental Diseases and Crime," 11–12.

37. Ibid., 11–12, 16.

38. "Chapter 34: An Act to Provide for the Sterilization of the Mentally Defective and Feeble-minded Inmates of Charitable and Penal Institutions of the the [*sic*] State of North Carolina," *Public Laws and Resolutions of the State of North Carolina* (1929), 28–29; North Carolina State Board of Charities and Public Welfare, *Biennial Report, 1920–1922*, 46–48. On sterilization in North Carolina, see Moya Woodside, *Sterilization in North Carolina: A Sociological and Psychological Study* (Chapel Hill: Univ. of North Carolina Press, 1950); Schoen, *Choice and Coercion*; and Castles, "Quiet Eugenics," 849–878.

39. Currell and Cogdell, eds., *Popular Eugenics*, 148–149.

40. North Carolina State Board of Charities and Public Welfare, *Biennial Report, 1926–1928*, 101.

5. A MODERN GIRL'S DILEMMA

1. Lewis, "Samarcand Arson Case, Carthage, N.C., May 18, 1931," in folder "Samarcand Arson Case," 1–3, Lewis Papers.

2. Cahn, *Sexual Reckonings*, 5, 10, 12, 14, 23, 45; Alexander, *The "Girl Problem,"* 150–151.

3. Lewis, "Samarcand Arson Case, Carthage, N.C., May 18, 1931," Lewis Papers.

4. Susan M. Cahn, *Sexual Reckonings: Southern Girls in a Troubling Age* (Cambridge: Harvard Univ. Press, 2007), 5–10, 18. Letters and complaints reveal that parents committed daughters who were raped when it seemed there was no other way to avoid scandal or bring the perpetrator to justice. See "Samarcand Manor: Investigations of Complaints, etc. 1926–1934," Institutions and Corrections, State Charitable, Penal, and Correctional Institutions, Records of the State Board of Public Welfare, box 165, North Carolina State Archives, Raleigh, North Carolina.

5. Statement of Viola Sistare, RN, March 21, 1931, in Lewis, "Samarcand Arson Case, Carthage, N.C., May 18, 1931," in folder "Samarcand Arson Case," 1, 3–5, Lewis Papers; Statement of Agnes MacNaughton, state investigation titled "Part II," in "Samarcand Manor, 1931," box 164, State Board of Charities and Public Welfare, Institutions and Corrections, State Charitable, Penal and Correctional Institutions, Samarcand, State Archives of North Carolina, Raleigh, North Carolina, 74. In 1926, the board of directors investigated a report by parents that their daughter was confined in a strong room for four months and whipped cruelly by thirteen girls with the knowledge and consent of the matron. The investigation verified the bruises on the child, who said she was whipped five times since she had been in Samarcand "with some kind of a plaited reed or lash." One matron claimed the girl had pinched herself to cause the bruises. The case was closed without action. See "Samarcand Manor, June 29, 1926," in "Samarcand Manor: Investigations of Complaints, etc. 1926–1934," Institutions and Corrections, State Charitable, Penal and Correctional Institutions, Records of the State Board of Public Welfare, box 165, North Carolina State Archives, Raleigh, North Carolina.

6. Statement of Viola Sistare, RN, March 21, 1931, in Lewis, "Samarcand Arson Case, Carthage, N.C., May 18, 1931," in folder "Samarcand Arson Case," 1, 3–5, Lewis Papers. See also

"Bessie Bishop to 'To Whom It May Concern,'" April 1, 1931, in "Samarcand Manor, 1931," box 164, State Board of Charities and Public Welfare, Institutions and Corrections, State Charitable, Penal and Correctional Institutions, Samarcand, State Archives of North Carolina, Raleigh, North Carolina, 33.

7. Sims, *The Power of Femininity in the New South,* 1–2, 30, 40. See also Scott, *The Southern Lady*; and Wheeler, *New Women of the New South*, 17–19; Hall, *Revolt Against Chivalry*; Hall, "'The Mind That Burns in Each Body,'" in Snitow, Stansell, and Thompson, eds., *Powers of Desire*; and Gilmore, *Gender and Jim Crow*; and Elna C. Green, *Southern Strategies: Southern Women and the Woman Suffrage Question* (Chapel Hill: Univ. of North Carolina Press, 1997).

8. Sims, *The Power of Femininity in the New South,* 51–58; Gilmore, *Gender and Jim Crow,* 31–32; Cahn, *Sexual Reckonings,* 13, 14.

9. Sims, *The Power of Femininity in the New South,* 157; Wheeler, *New Women of the New South,* xv, 38–40; Sara M. Evans, *Born for Liberty: A History of Women in America* (New York: Free Press, 1989), 161, 175–179, 184; Cahn, "Spirited Youth or Fiends Incarnate," 160. See also Cahn, *Sexual Reckonings,* 30, 51.

10. F. Scott Fitzgerald, *Flappers and Philosophers* (New York: Scribner and Sons, 1920), 17–19, 23; Cahn, *Sexual Reckonings,* 5, 16–18.

11. Fitzgerald, *Flappers and Philosophers,* 36–38.

12. Ibid., chapter 5.

13. Joshua Zeitz, *Flapper: A Madcap Story of Sex, Style, Celebrity, and the Women Who Made American Modern* (New York: Crown Publishers, 1982), 29–30, 39, quotes on page 211, 233; Catherine Gourley, *Flappers and the New American Woman* (Minneapolis: Twenty-First Century Books, 2008), 65–67; Evans, *Born for Liberty,* 178–179. See also Angela J. Latham, *Posing a Threat: Flappers, Chorus Girls, and Other Brazen Performers of the American 1920s* (Hanover, N.H.: Wesleyan Univ. Press, 2000); and Paula S. Fass, *The Damned and the Beautiful: American Youth in the 1920's* (New York: Oxford Univ. Press, 1977).

14. Gourley, *Flappers and the New American Woman,* 72–73, 83, 84–85, 93, 102; Zeitz, *Flapper,* 173, 196–203.

15. Lux Toilet Soap advertisement, *Raleigh (N.C.) News and Observer,* January 23, 1929, 2, col. 7; Report of Ruth Bloodgood, March 7–10, 1935, State Board of Public Welfare, State Charitable, Penal, and Correctional Institutions, in "Samarcand Manor, 1918–1924," box 164, State Board of Charities and Public Welfare, Institutions and Corrections, State Charitable, Penal and Correctional Institutions, Samarcand, State Archives of North Carolina, Raleigh, North Carolina

16. "Bevy of Charming North Carolina Brides and Brides-to-be of the Mid Winter Season," *Raleigh (N.C.) News and Observer,* January 27, 1929, 1 (Society section); January 20, 1929, 1 (Society section); February 28, 1929, 6, col. 5. See also "Bride of the Winter Season" and "Attractive Raleigh Bride," *Raleigh (N.C.) News and Observer,* February 27, 1929, 6, col. 4.

17. "Raleigh Girls Selected for Parts in Local Film," *Raleigh (N.C.) News and Observer,* February 26, 1929, 2, col. 5; advertisement of the Boylan and Pearce Company, *Raleigh (N.C.) News and Observer,* January 24, 1929, 1 (Local section).

18. "Arrest Girl Found Wearing Men's Garb," *Raleigh (N.C.) News and Observer,* January 25, 1929:, 9, col. 3.

19. "Four Girls Check out of Asheboro," *Raleigh (N.C.) News and Observer,* January 21, 1929, 2, col. 4; and "Martin County Girl Is on Missing List," *Raleigh (N.C.) News and Observer,* February 17, 1929, 4, col. 2.

20. Kunzel, *Fallen Women, Problem Girls,* 147; Lena B. Ladu and K. C. Garrison, "A Study of Emotional Instability and Intelligence of Women in the Penal Institutions of North Carolina," *Social Forces* 10, no. 2 (December 1931): 210–212, quote on page 212.

21. Ladu and Garrison, "A Study of Emotional Instability and Intelligence of Women in the Penal Institutions of North Carolina," 212–215, quote on page 213.

22. "Martin County Girl Is on Missing List," *Raleigh (N.C.) News and Observer,* February 17, 1929, 4, col. 2; "Four Girls Check Out of Asheboro," *Raleigh (N.C.) News and Observer,* January 21, 1929, 2, col. 4; "Girl Leaves Home with Man; Search Started," *Raleigh (N.C.) News and Observer,* February 21, 1929, 8, col. 3; "Four Missing Girls All Located Now," *Raleigh (N.C.) News and Observer,* January 24, 1929, 3, col. 5; "Girl Gone, Writes She Has Married," *Raleigh (N.C.) News and Observer,* January 26, 1929, 2, col. 6; "Two More Girls Set Out to See World," *Raleigh (N.C.) News and Observer,* January 28, 1929, 2, col. 6.

23. "Press Search for Two Missing Girls," *Raleigh (N.C.) News and Observer,* February 5, 1929, 1, col. 6; "Low Grades in School Caused Girls to Leave," *Raleigh (N.C.) News and Observer,* February 7, 1929, 1, col. 6; "Greensboro Girls Found in Florida," *Raleigh (N.C.) News and Observer,* February 6, 1929, 12, col. 5; "Girl Gone, Writes She Has Married," *Raleigh (N.C.) News and Observer,* January 26, 1929, 2, col. 6; "Girl Leaves Home with Man; Search Started," *Raleigh (N.C.) News and Observer,* February 21, 1929, 8, col. 5.

24. "Four Girls Check Out of Asheboro," *Raleigh (N.C.) News and Observer,* January 21, 1929, 2, col. 4; political cartoon, "English J. P. Tames Flaming Youth with Hair Brush," *Raleigh (N.C.) News and Observer,* February 7, 1929, 4, col. 3.

25. "Wilmington Hunters Are Believed Dead in Swamp," *Raleigh (N.C.) News and Observer,* January 21, 1929, 1, col. 6; "No Trace Found of Missing Boys," *Raleigh (N.C.) News and Observer,* January 24, 1929, 8, col. 5; "Scent Foul Play in Disappearance," *Raleigh (N.C.) News and Observer,* January 25, 1929, 12, col. 6; "Runaway Youngsters Are Taken Back Home," *Raleigh (N.C.) News and Observer,* March 1929, 7, col. 2.

26. "Wilmington Hunters Are Believed Dead in Swamp," *Raleigh (N.C.) News and Observer,* January 21, 1929, 1, col. 6; "Still No Clue On 4 Missing Hunters," *Raleigh (N.C.) News and Observer,* January 28, 1929, 10, col. 1; "Body of Another Hunter Found," *Raleigh (N.C.) News and Observer,* February 11, 1929, 4, col. 5; "Recover Body of Jesse Huggins," *Raleigh (N.C.) News and Observer,* February 21, 1929, 12, col. 3; "Body of Fourth of Missing Party Found," *Raleigh (N.C.) News and Observer,* February 24, 1929, 2, col. 3.

27. "Kidnapped Girl Is Found in Box Car Bound by Rope," *Raleigh (N.C.) News and Observer,* April 10, 1931, 1, col. 6; "Police Make No Headway in Kidnapping Mystery," *Raleigh (N.C.) News and Observer,* April 11, 1931, 2, col. 1. See also "Five Men Try to Carry Child Off," *Raleigh (N.C.) News and Observer,* February 12, 1929, 1, col. 2.

28. "Confesses 'Kidnapping' by Men Was Only Farce," *Raleigh (N.C.) News and Observer,* April 12, 1931, 2, col. 1.

29. Georganne Scheiner, *Signifying Female Adolescence: Film Representations and Fans, 1920–1950* (Westport, Conn.: Praeger, 2000), 23, 35.

30. Cahn, "Spirited Youth or Fiends Incarnate," 160; Scheiner, *Signifying Female Adolescence,* 6–9; Alexander, *The "Girl Problem"*; Jacquelyn Dowd Hall: "Disorderly Women: Gender and Labor Militancy in the Appalachian South," *Journal of American History* 73, no. 2 (September 1986): 377; Odem, *Delinquent Daughters,* 1–5; Constance Nathanson, *Dangerous Passage: The Social Control of Sexuality in Women's Adolescence* (Philadelphia: Temple Univ. Press, 1991); Peiss, *Cheap Amuse-*

ments; G. Stanley Hall, *Adolescence: Its Psychology and Its Relation to Physiology, Anthropology, Sociology, Sex, Crime, Religion and Education,* 2 vols. (New York: Appleton, 1904), 551–647. See also Cahn, *Sexual Reckonings*; Kunzel, *Fallen Women, Problem Girls.*

31. Lewis, "Samarcand Arson Case, Carthage, N.C., May 18, 1931," in folder "Samarcand Arson Case," Lewis Papers.

32. Ibid.

33. Hall, "Disorderly Women," 356, 360, 377. See also John Salmond, *The General Textile Strike of 1934: From Maine to Alabama* (Columbia: Univ. of Missouri Press, 2002).

34. Ibid.; statement of Agnes MacNaughton, state investigation titled "Part II," in "Samarcand Manor, 1931," box 164, State Board of Charities and Public Welfare, Institutions and Corrections, State Charitable, Penal and Correctional Institutions, Samarcand, State Archives of North Carolina, Raleigh, North Carolina, 39; Cahn, *Sexual Reckonings,* 18, 25, 48, 148.

35. Victoria Byerly and Cletus E. Daniel, *Hard Times Cotton Mill Girls: Personal Histories of Womanhood and Poverty in the South* (Ithaca: Cornell Univ. Press, 1986), 14, 17, 21, 23.

36. Hall, "Disorderly Women," 372–375.

37. Statement of Agnes MacNaughton, state investigation titled "Part II," in "Samarcand Manor, 1931," box 164, State Board of Charities and Public Welfare, Institutions and Corrections, State Charitable, Penal and Correctional Institutions, Samarcand, State Archives of North Carolina, Raleigh, North Carolina, 76–77.

38. Lewis, "Samarcand Arson Case, Carthage, N.C., May 18, 1931," in folder "Samarcand Arson Case," 1–3, Lewis Papers.

6. NOT PENITENT YET

1. "Samarcand Girls Riot in Moore County Jail," April 30, 1931, and "Samarcand Girls Go On Another Jail Rampage," May 1, 1931, *Raleigh (N.C.) News and Observer,* both in Lewis, "Samarcand Arson Case, Carthage, N.C., May 18, 1931," in folder "Samarcand Arson Case," Lewis Papers. For article-length accounts of the trial, see Cahn, "Spirited Youth or Fiends Incarnate," 152–180; Wertheimer, Luskey, et al., "Escape of the Match-Strikers," 435–460; Bickford, "Imperial Modernity, National Identity and Capital Punishment in the Samarcand Arson Case, 1931," 437–460.

2. "Miss Lewis to Appear in Arson Case," *Raleigh (N.C.) News and Observer,* April 25, 1931, in Lewis, "Samarcand Arson Case, Carthage, N.C., May 18, 1931," in folder "Samarcand Arson Case," Lewis Papers; Nell Battle Lewis, ed., *Capital Punishment in North Carolina: Special Bulletin Number 10* (Raleigh: North Carolina State Board of Charities and Public Welfare, 1929). See also Cahn, "Spirited Youth or Fiends Incarnate," 152–180; Cahn, *Sexual Reckonings.* See also McRae, "To Save a Home," 3, and biographical description, Nell Battle Lewis Collection, Private Collections index, North Carolina State Archives, Raleigh, North Carolina. See also "Says Insane Are Killed by State," *Raleigh (N.C.) News and Observer,* February 28, 1929, 8, col. 1.

3. "Virginia Hayes," in Lewis, "Samarcand Arson Case, Carthage, N.C., May 18, 1931," in folder "Samarcand Arson Case," Lewis Papers.

4. "Ms. MacNaughton," in "Information Collected by NBL at Samarcand," in Lewis, "Samarcand Arson Case, Carthage, N.C., May 18, 1931," in folder "Samarcand Arson Case," Lewis Papers.

5. Ibid.

6. "Miss Moore," "Miss Ross," and "Miss Stott," in "Information Collected by NBL at Samarcand," in Lewis, "Samarcand Arson Case, Carthage, N.C., May 18, 1931," in folder "Samarcand Arson Case," Lewis Papers.

7. See articles 14 and 15 in "Ch. 82: Crime and Punishments," *The North Carolina Code of 1939* (Charlottesville, Va: Michie Company, Law Publishers, 1939), 1677–1679.

8. Jeff Welty, "The Death Penalty in North Carolina: History and Overview," April 2012, www .unc.edu/~fbaum/teaching/articles/Welty-DP-overview.pdf (accessed January 2, 2014); Lewis, *Capital Punishment in North Carolina*, 11–13, 15–17; Kathleen O'Shea, *Women and the Death Penalty in the United States, 1900–1998* (Westport, Conn.: Praeger, 1999), 23.

9. Victor Streib, "Death Penalty for Female Offenders," *University of Cincinnati Law Review* 58 (1990) 845, 855, 866; Timothy V. Kaufman-Osborn, "Symposium on Gender and the Death Penalty," *Signs: Journal of Women in Culture and Society* 24, no. 4 (summer 1999): 1097; O'Shea, *Women and the Death Penalty,* 23.

10. On state executions as substitutions for lynching, see Charles D. Phillips, "Social Structure and Social Control: Modeling the Discriminatory Execution of Blacks in Georgia and North Carolina, 1925–1935," *Social Forces* 65, no. 2 (December 1986): 458–475, 458–460; Charles D. Phillips, "Exploring Relations among Forms of Social Control: The Lynching and Execution of Blacks in North Carolina, 1889–1918," *Law and Society Review* 21, no. 3 (1987): 361–374; James W. Clarke, "Without Fear or Shame: Lynching, Capital Punishment and the Subculture of Violence in the American South," *British Journal of Political Science* 28, no. 2 (April 1998): 269–289. Arguing against the substitution model are E. M. Beck, James L. Massey, and Stewart E. Tolnay, "The Gallows, the Mob, and the Vote: Lethal Sanctioning of Blacks in North Carolina and Georgia, 1882 to 1930," *Law and Society Review* 23, no. 2 (1989): 317–331; W. Fitzhugh Brundage, *Lynching in the New South: Georgia and Virginia, 1880–1930* (Urbana: Univ. of Illinois Press, 1993), quote on page 256; Margaret Vandiver, *Lethal Punishment: Lynchings and Legal Executions in the South* (New Brunswick, N.J.: Rutgers Univ. Press, 2005). On lynching and white supremacy, see Hall, *Revolt Against Chivalry*; and Gilmore, *Gender and Jim Crow.*

11. "Says She Caused $200,000 Blaze," *Raleigh (N.C.) News and Observer,* May 24, 1931, 10, col. 3; "Accuse Girls of Burning Buildings at Samarcand," *Raleigh (N.C.) News and Observer,* March 17, 1931, 1, col. 3.

12. Jacqueline Dowd Hall et al., *Like a Family: The Making of a Southern Cotton Mill World* (Chapel Hill: Univ. of North Carolina Press, 2000), 43, 114, 214, 216–217; Salmond, *The General Textile Strike of 1934,* 3–4, 8–10. See also John A. Salmond, "The Burlington Dynamite Plot: The 1934 Textile Strike and Its Aftermath in Burlington, North Carolina," *North Carolina Historical Review* 75, no. 4 (October 1998): 398–434.

13. Salmond, *The General Textile Strike,* 9; Hall et al., *Like a Family,* 214.

14. Hall et al., *Like a Family,* 217.

15. John A. Salmond, *Gastonia, 1929: The Story of the Loray Mill Strike* (Chapel Hill: Univ. of North Carolina Press, 1995), 20, 25 (picture), 34, 50–52.

16. Ibid., 62.

17. Ibid., 127–131, quote on page 130; "Incidentally," *Raleigh (N.C.) News and Observer,* September 22, 1929.

18. Salmond, *Gastonia, 1929,* 26; Robert Weldon Whalen, "'*Like Fire in Broom Straw': Southern Journalism and the Textile Strikes of 1929–1931*" (Westport, Conn.: Greenwood Press, 2001), 31–32;

Gastonia (N.C.) Daily Gazette, April 3, 1929, 1, 7, and April 4, 1929, 1, 7; "Clark Scouts Importance of Youth's Strike at Gastonia: Publisher Declares Communistic Program of Free-Love and Inter-Racial Intercourse Misses Fire," *Charlotte (N.C.) Observer,* April 6, 1929, 1.

19. Salmond, *Gastonia, 1929,* 30–32; Robert Weldon Whalen, "Recollecting the Cotton Mill Wars: Proletarian Literature of the 1929–1931 Southern Textile Strikes," *North Carolina Historical Review* 75, no. 4 (October 1998): 388–394; Hall, "Disorderly Women," 354–382; *Gastonia (N.C.) Daily Gazette,* April 18, 1929, April 25, 1929.

20. Cahn, "Spirited Youth or Fiends Incarnate," 154, quote on page 155. On court depictions of poor women as disorderly, see Victoria E. Bynum, *Unruly Women: The Politics of Social and Sexual Control in the Old South* (Chapel Hill: Univ. of North Carolina Press, 1992); Bickford, "Imperial Modernity, National Identity, and Capital Punishment in the Samarcand Arson Case, 1931," 442–43; and Cahn, *Sexual Reckonings.*

21. Crista DeLuzio, *Female Adolescence in American Scientific Thought, 1830–1930* (Baltimore: Johns Hopkins Univ. Press, 2007), 90–96.

22. Ibid., 109–114, quote on page 112.

23. "Resolution No. 53: A Joint Resolution Relative to the Safe Keeping of the Girls, Formerly Inmates of Samarcand, Now Under Indictment for Firing the Building of that Institution," *Public Laws and Resolutions of the State of North Carolina* (1931), 808.

24. Carolyn L. Reynolds, "Rebellious Girls Set Their Bunks Afire to Get Thrill," *Raleigh (N.C.) News and Observer,* May 2, 1931, in folder "Samarcand Arson Case," Lewis Papers.

25. Ibid.

26. Ibid.

27. Ibid.

28. See Kate Burr Johnson, "State Welfare Program as It Relates to Mental Diseases and Crime," speech to the Medical Society, Lenoir, North Carolina, September 1927, "Speeches, Reports and Miscellaneous," 91.1.c, Kate Ancrum Burr Johnson Papers, 14; Erskine Caldwell, *Tobacco Road* (Cutchogue, N.Y.: Buccaneer Books, 1932); and Erskine Caldwell, *God's Little Acre* (New York: Viking, 1933).

29. "Samarcand Girls Are Quietly Awaiting News of Their Fate," *Moore County (N.C.) News,* ca. May 7, 1931, in folder "Samarcand Arson Case," Lewis Papers.

30. "Reform Cshool [sic] Boys Are Tried," *Raleigh (N.C.) News and Observer,* April 22, 1931, and "Young Firebug Gets One Year," *Charlotte (N.C.) Observer,* April 26, 1931, both in folder "Samarcand Arson Case," Lewis Papers.

31. "Defense Holds Samarcand Girls Victims State Neglect," *Raleigh (N.C.) News and Observer,* May 20, 1931, in folder "Samarcand Arson Case," Lewis Papers.

32. Ibid.

33. Ibid.

34. Ibid., "Information Collected by NBL at Samarcand," in Lewis, "Samarcand Arson Case, Carthage, N.C., May 18, 1931," in folder "Samarcand Arson Case," Lewis Papers.

35. "Defense Holds Samarcand Girls Victims State Neglect," *Raleigh (N.C.) News and Observer,* May 20, 1931, in folder "Samarcand Arson Case," Lewis Papers.

36. Ibid.; State Investigation of Samarcand Manor, State Industrial Training School for Girls, 1931, in "Samarcand Manor, 1931," box 164, State Board of Charities and Public Welfare, Institutions and Corrections, State Charitable, Penal and Correctional Institutions, Samarcand, State

Archives of North Carolina, Raleigh, North Carolina; "Statement of Lottie Mitchem," March 27, 1931, and "Defense Testimony: Miss Lottie Mitchem," in Lewis, "Samarcand Arson Case, Carthage, N.C., May 18, 1931," both in folder "Samarcand Arson Case," Lewis Papers.

37. "Defense Holds Samarcand Girls Victims State Neglect," *Raleigh (N.C.) News and Observer,* May 20, 1931, in folder "Samarcand Arson Case," Lewis Papers. Susan Cahn argues that Stiles couched her innocence in the language of childhood adolescence. Though she used the rhetoric of revolutionary patriotism, Cahn explains, her language was more tentative than bold. See Cahn, "Spirited Youth or Fiends Incarnate," 170–171. Pearl Stiles's letter is located in folder "Samarcand Manor, 1931," box 164, State Board of Charities and Public Welfare, Institutions and Corrections, State Charitable, Penal and Correctional Institutions, Samarcand, State Archives of North Carolina, Raleigh, North Carolina.

38. "Defense Holds Samarcand Girls Victims State Neglect," *Raleigh (N.C.) News and Observer,* May 20, 1931, in folder "Samarcand Arson Case," Lewis Papers.

39. Emily Seibel, "Nell Battle Lewis in the Progressive Era: Prison and Capital Punishment Reform" (Master's thesis, University of North Carolina at Asheville, 2009), 15–16; Lewis, "Capital Punishment in North Carolina," 172; Nell Battle Lewis, "Incidentally," *Raleigh (N.C.) News and Observer,* August 4, 1935.

40. "Defense Holds Samarcand Girls Victims State Neglect," *Raleigh (N.C.) News and Observer,* May 20, 1931, in folder "Samarcand Arson Case," Lewis Papers; Harry W. Crane to Nell Battle Lewis, May 13, 1931, letter included in Lewis, "Samarcand Arson Case, Carthage, N.C., May 18, 1931," in folder "Samarcand Arson Case," Lewis Papers.

41. "Twelve Samarcand Girls Get State Prison Terms," *Raleigh (N.C.) News and Observer,* May 21, 1931, in folder "Samarcand Arson Case," Lewis Papers.

42. Ibid.

43. Ibid.

7. CLASSIFYING "SUBNORMAL"

1. "Bad Conditions at Samarand," *Chapel Hill (N.C.) Weekly,* May 29, 1931; "North Carolina Fails," n.d., *Rocky Mount (N.C.) Telegram,* all in "Samarcand Arson Case," Lewis Papers.

2. Report summary, in "Samarcand Manor, 1925–32," box 164, State Board of Charities and Public Welfare, Institutions and Corrections, State Charitable, Penal and Correctional Institutions, Samarcand, State Archives of North Carolina, Raleigh, North Carolina.

3. Statement of Viola Sistare, March 21, 1931; statement of Arline James, March 21, 1931; statement of Grace Henslee, circa March 27, 1931, Georgia Piland to Mrs. W. T. [Annie Kizer] Bost, March 3, 1931; statement of Georgia Piland, March 27, 1931; statement of Lottie Mitchem, March 27, 1931, all in "Part I," Report of Investigation at Samarcand Manor, in "Samarcand Manor, 1931," box 164, State Board of Charities and Public Welfare, Institutions and Corrections, State Charitable, Penal and Correctional Institutions, Samarcand, State Archives of North Carolina, Raleigh, North Carolina. Copies of these statements also appear in the Nell Battle Lewis Papers, "Samarcand Arson Case," box 255.29.

4. Georgia Piland to Mrs. W. T. [Annie Kizer] Bost, March 3, 1931; statement of Georgia Piland, March 27, 1931; statement of Lottie Mitchem, March 27, 1931; "Personnel Resigned 1929,

1930, and to April 1931" and "Current Personnel March 30, 1931," all in "Part I," Report of Investigation at Samarcand Manor, in "Samarcand Manor, 1931," box 164, State Board of Charities and Public Welfare, Institutions and Corrections, State Charitable, Penal and Correctional Institutions, Samarcand, State Archives of North Carolina, Raleigh, North Carolina.

5. Statement of Viola Sistare, March 21, 1931; statement of Lottie Mitchem, March 27, 1931, both in "Part I," Report of Investigation at Samarcand Manor, in "Samarcand Manor, 1931," box 164, State Board of Charities and Public Welfare, Institutions and Corrections, State Charitable, Penal and Correctional Institutions, Samarcand, State Archives of North Carolina, Raleigh, North Carolina. Copies of these statements also appear in the Lewis Papers, "Samarcand Arson Case"; "Smoke Signals From Samarcand," in "Samarcand Arson Case," "Letters to the Editor," Lewis Papers, box 255.29; "Letter to the Editor," *Fayetteville (N.C.) Observer*, May 4, 1931.

6. "Personnel Questionnaire," statement of Mary Lee Hunt, April 1, 1931, in "Part I," Report of Investigation at Samarcand Manor, in "Samarcand Manor, 1931," box 164, State Board of Charities and Public Welfare, Institutions and Corrections, State Charitable, Penal and Correctional Institutions, Samarcand, State Archives of North Carolina, Raleigh, North Carolina, quote by girls, 74.

7. "Personnel Questionnaire," statement of Bessie Bishop, April 1, 1931, in "Part I," Report of Investigation at Samarcand Manor, in "Samarcand Manor, 1931," box 164, State Board of Charities and Public Welfare, Institutions and Corrections, State Charitable, Penal and Correctional Institutions, Samarcand, State Archives of North Carolina, Raleigh, North Carolina.

8. "Personnel Questionnaire," statement of Maude Moore, April 1, 1931, and statement of J. H. Bodenheimer, April 2, 1931, in "Part I," Report of Investigation at Samarcand Manor, in "Samarcand Manor, 1931," box 164, State Board of Charities and Public Welfare, Institutions and Corrections, State Charitable, Penal and Correctional Institutions, Samarcand, State Archives of North Carolina, Raleigh, North Carolina.

9. Karin Zipf, "No Longer under Cover(ture): Marriage, Divorce, and Gender in the 1868 Constitutional Convention," in *North Carolinians in the Era of the Civil War and Reconstruction*, ed. Paul D. Escott (Chapel Hill: Univ. of North Carolina Press, 2008), 197.

10. Report of Investigation at Samarcand Manor, in "Samarcand Manor, 1931," box 164, State Board of Charities and Public Welfare, Institutions and Corrections, State Charitable, Penal and Correctional Institutions, Samarcand, State Archives of North Carolina, Raleigh, North Carolina, page 2.

11. Ibid., 36–37, 63.

12. Number 12 listed no demerits but required a "call down," most likely a visit to the Discipline Officer, Crenshaw. Report of Investigation at Samarcand Manor, in "Samarcand Manor, 1931," box 164, State Board of Charities and Public Welfare, Institutions and Corrections, State Charitable, Penal and Correctional Institutions, Samarcand, State Archives of North Carolina, Raleigh, North Carolina, pages 40–41.

13. Ibid., 82.

14. Ibid., 85–86.

15. Ibid.

16. *Raleigh (N.C.) News and Observer,* May 31, 1931, "Board of Samarcand Bans Whipping Girl Inmates," clipping in "Samarcand Arson Case," Lewis Papers.

17. Minnie R. Kimball to Mrs. T. W. Bost, September 27, 1932, in "Samarcand Manor, 1932–1937," box 164, State Board of Charities and Public Welfare, Institutions and Corrections, State

Charitable, Penal and Correctional Institutions, Samarcand, State Archives of North Carolina, Raleigh, North Carolina.

18. Director, Division of Institutions, to Miss Agnes MacNaughton, October 4, 1932, and Minnie R. Kimball to R. Eugene Brown, November 3, 1932, in "Samarcand Manor, 1932–1937," box 164, State Board of Charities and Public Welfare, Institutions and Corrections, State Charitable, Penal and Correctional Institutions, Samarcand, State Archives of North Carolina, Raleigh, North Carolina.

19. Mrs. W. T. [Annie Kizer] Bost to Miss Helene Ulrich, March 12, 1934; Minnie Ross Kimball to Mrs. W. T. [Annie Kizer] Bost, March 1, 1934; Helene Ulrich to Mrs. W. T. [Annie Kizer] Bost, March 9, 1934; Mrs. W. T. [Annie Kizer] Bost to Dr. W. A. Stanbury, June 2, 1934; Mrs. W. T. [Annie Kizer] Bost to Mrs. J. R. Page, August 13, 1934, all in "Samarcand Manor, 1932–1937," box 164, State Board of Charities and Public Welfare, Institutions and Corrections, State Charitable, Penal and Correctional Institutions, Samarcand, State Archives of North Carolina, Raleigh, North Carolina; "Nursing News and Announcements," *American Journal of Nursing* 12, no. 9 (June 1912): 752; *Proceedings of the Quarterly Conference: New Jersey State Board of Control of Institutions and Agencies, 1922* (Memphis: General Books, 2010).

20. Herbert O'Keef, "Samarcand Is Place of Multitudinous Problems: Institution Undertakes Solution of Human Puzzles When Home, School, Church and Community Have Failed," *Raleigh (N.C.) News and Observer*, April 8, 1934, 1M—2M.

21. L. L. Boyd to Roy E. Brown, April 9, 1934, and Kate Capehart to Mrs. W. T. [Annie Kizer] Bost, May 11, 1934, both in "Samarcand Manor, 1932–1937," box 164, State Board of Charities and Public Welfare, Institutions and Corrections, State Charitable, Penal and Correctional Institutions, Samarcand, State Archives of North Carolina, Raleigh, North Carolina.

22. Mrs. W. T. [Annie Kizer] Bost to Kate Capehart, May 16, 1934, "Samarcand Manor, 1932–1937," box 164, State Board of Charities and Public Welfare, Institutions and Corrections, State Charitable, Penal and Correctional Institutions, Samarcand, State Archives of North Carolina, Raleigh, North Carolina; O'Keef, "Samarcand Is Place of Multitudinous Problems," 2M.

23. Lily E. Mitchell, Director of Child Welfare, to Agnes B. MacNaughton, December 13, 1933; Lily E. Mitchell, to J. P. Marsh, December 22, 1933; Lily E. Mitchell to Agnes B. MacNaughton, December 22, 1933; Grace M. Robson to Dr. Delia Dixon-Carroll, May 7, 1934, all in "Samarcand Manor, 1932–1937," box 164, State Board of Charities and Public Welfare, Institutions and Corrections, State Charitable, Penal and Correctional Institutions, Samarcand, State Archives of North Carolina, Raleigh, North Carolina.

24. "Pre-Parole Review of 'Eliza,'" January 15, 1935, in "Samarcand Manor, 1932–1937," box 164, State Board of Charities and Public Welfare, Institutions and Corrections, State Charitable, Penal and Correctional Institutions, Samarcand, State Archives of North Carolina, Raleigh, North Carolina.

25. Dr. J. P. Bowen, "Standards for the Care of Venereal Diseases at Samarcand Manor, Eagle Springs, N.C. submitted by Dr. J. P. Bowen, Physician in Charge of Medical Department," October 31, 1934, Samarcand Manor, Archives, Eagle Springs, North Carolina.

26. Ruth Bloodgood, Report on the State Home Industrial School for Girls, March 7–10, 1935, in "Samarcand Manor, 1932–1937," box 164, State Board of Charities and Public Welfare, Institutions and Corrections, State Charitable, Penal and Correctional Institutions, Samarcand, State Archives of North Carolina, Raleigh, North Carolina, page 9.

27. Grace M. Robson, "Institutional Treatment," presented to the Fifteenth Annual Public Welfare Institute, Chapel Hill, North Carolina, 1933, in "Samarcand Manor, 1932–1937," box 164, State Board of Charities and Public Welfare, Institutions and Corrections, State Charitable, Penal and Correctional Institutions, Samarcand, State Archives of North Carolina, Raleigh, North Carolina; Grace M. Robson, "Samarcand Manor," in "First Juvenile Delinquent Training School in North Carolina Was Opened in 1909," *Public Welfare News* 7, no. 1 (March 1944): 26–27.

28. Grace M. Robson, "Institutional Treatment," presented to the Fifteenth Annual Public Welfare Institute, Chapel Hill, North Carolina, 1933, in "Samarcand Manor, 1932–1937," box 164, State Board of Charities and Public Welfare, Institutions and Corrections, State Charitable, Penal and Correctional Institutions, Samarcand, State Archives of North Carolina, Raleigh, North Carolina.

29. Grace M. Robson, "Institutional Treatment," presented to the Fifteenth Annual Public Welfare Institute, Chapel Hill, North Carolina, 1933, in "Samarcand Manor, 1932–1937," box 164, State Board of Charities and Public Welfare, Institutions and Corrections, State Charitable, Penal and Correctional Institutions, Samarcand, State Archives of North Carolina, Raleigh, North Carolina; Grace M. Robson, "Samarcand Manor," *Public Welfare News* 7, no. 1 (March 1944): 26–27.

30. Michael A. Rembis, "'I Ain't Been Reading While on Parole': Experts, Mental Tests, and Eugenic Commitment Law in Illinois, 1890–1940," *History of Psychology* 7, no. 3 (August 2004): 227–228, 233–236. See also Michael A. Rembis, *Defining Deviance: Sex, Science, and Delinquent Girls, 1890–1960* (Urbana: Univ. of Illinois Press, 2013).

31. Grace M. Robson, "Institutional Treatment," presented to the Fifteenth Annual Public Welfare Institute, Chapel Hill, North Carolina, 1933, in "Samarcand Manor, 1932–1937," box 164, State Board of Charities and Public Welfare, Institutions and Corrections, State Charitable, Penal and Correctional Institutions, Samarcand, State Archives of North Carolina, Raleigh, North Carolina.

32. Cahn, *Sexual Reckonings*, 157.

33. Ibid., 158–159.

34. "Chapter 281: An Act to Benefit the Moral, Mental, or Physical Conditions of Inmates of Penal and Charitable Institutions," *Public Laws and Resolutions of the State of North Carolina* (1919), 504; "Chapter 34: An Act to Provide for the Sterilization of the Mentally Defective and Feeble-Minded Inmates of Charitable and Penal Institutions of the [*sic*] State of North Carolina," *Public Laws and Resolutions of the State of North Carolina* (1929), 28–29; "Senate Favors Sterilization Bill," *Raleigh (N.C.) News and Observer*, February 8, 1929, 5, col. 1; "Sterilizing Bill Passes in Senate," *Raleigh (N.C.) News and Observer*, February 9, 1929, 1, col. 2; "Curb on Mental Defectives Now Provided by Law," *Raleigh (N.C.) News and Observer* August 17, 1929, 1, col. 1; "Eugenical Sterilization in North Carolina," Report of the Eugenics Board, 1935, in "State Board of Charities and Public Welfare, General Correspondence, 1933–1937," box 145, Governors Papers, J.C.B. Ehringhaus, State Archives of North Carolina, Raleigh, North Carolina, 7; *Brewer v. Valk* (1933) 204 NC 186. The quote "mechanically cumbersome" comes from Schoen, *Choice and Coercion*, 81–82.

35. "Chapter 224: An Act to Amend Chapter 34 of the Public Laws of 1929 of North Carolina Relating to the Sterilization of Persons Mentally Defective," *Public Laws and Resolutions of the State of North Carolina* (1933), 345–352; *Raleigh (N.C.) News and Observer*, December 30, 1934; Report of the Eugenics Board, 1935, in "State Board of Charities and Public Welfare, General Correspondence, 1933–1937," box 145, Governors Papers, J.C.B. Ehringhaus, State Archives of North Carolina, Raleigh, North Carolina, 7; Schoen, *Choice and Coercion*, 82.

36. Robson, "Samarcand Manor," 26–27.

37. Report of the Eugenics Board, 1935, in "State Board of Charities and Public Welfare, General Correspondence, 1933–1937," box 145, Governors Papers, J.C.B. Ehringhaus, State Archives of North Carolina, Raleigh, North Carolina; Rembis, "'I Ain't Been Reading While on Parole,'" 233–236; Kuhl, *The Nazi Connection,* xvii–xviii, 39; Woodside, *Sterilization in North Carolina,* 23–24.

38. Woodside, *Sterilization in North Carolina,* 23–24.

39. "Pre-Parole Review of 'Ida,'" January 15, 1935, in "Samarcand Manor, 1932–1937," box 164, State Board of Charities and Public Welfare, Institutions and Corrections, State Charitable, Penal and Correctional Institutions, Samarcand, State Archives of North Carolina, Raleigh, North Carolina.

40. This data is compiled from the biennial reports, General File, 1933–1974, Eugenics Commission, North Carolina State Archives, Raleigh, North Carolina.

41. Woodside, *Sterilization in North Carolina,* xiii–xiv, 51–52, 164–165.

42. Ibid., 35–38.

43. "Chronological History: Re-Classification Report of 'Melba Hall,'" October 22, 1935, Samarcand Manor Scrapbook Collection, Eagle Springs, North Carolina.

44. Ibid.

45. Ibid.

46. Woodside, *Sterilization in North Carolina,* 99–100.

47. Ibid., 36–38, 71.

48. Ibid., 37, 126.

49. "Classification: October 19, 1934," "Classification: November 23, 1934," "Classification: December 7, 1934," and "Classification: December 21, 1934," all in Samarcand Manor Scrapbook Collection, Eagle Springs, North Carolina.

50. Ellen Winston to R. O. Heater, "April 29, 1953," County Director Letters, Records of the State Board of Charities and Public Welfare, RG 97, State Archives of North Carolina, Raleigh, North Carolina; Schoen, *Choice and Coercion,* 18–19, 103–112, 129–130; Johanna Schoen, "From the Footnotes to the Headlines: Sterilization Apologies and Their Lessons," *Sexuality Research and Social Policy* 3, no. 3 (September 2006): 7, 16–18.

8. "A MYSTERY TO ME"

1. "Investigation Samarcand Ordered," *Moore County (N.C.) News,* January 21, 1943.

2. Ibid.; "Armstrong Instructs Jury to Investigate Conditions of Samarcand Manor Girls," *The Pilot,* Southern Pines, North Carolina, January 22, 1943.

3. Drury B. Thompson, *Juvenile Courts in North Carolina* (Raleigh, N.C.: State Board of Public Welfare, 1951), 9, 11; *State Laws Governing Public Welfare Work in North Carolina* (Raleigh, N.C.: North Carolina State Board of Charities and Public Welfare, 1923) 20; *A Square Deal for the Child: Juvenile Courts in North Carolina* (Raleigh, N.C.: North Carolina State Board of Charities and Public Welfare, 1923), 3. These documents rest in the Records of the North Carolina Board of Public Welfare, box 33, series 93, Outer Banks History Center, North Carolina State Archives, Manteo, North Carolina.

4. L. Mara Dodge, *"Whores and Thieves of the Worst Kind": A Study of Women, Crime, and Prisons, 1835–2000* (DeKalb: Northern Illinois Univ. Press, 2006), 19–21; see also Nicole Hahn

Rafter, *Partial Justice: Women in State Prisons, 1800–1935* (Boston: Northeastern Univ. Press, 1985); Joanne Belknap, *The Invisible Woman: Gender, Crime, and Justice* (Belmont, Calif.: Wadsworth, 1996); Rebecca Ragsdale Lallier, "'A Place of Beginning Again': The North Carolina Industrial Farm Colony for Women, 1929–1947" (Master's thesis, University of North Carolina at Chapel Hill, 1990), 67–72, 92.

5. Wiley Britton Sanders and William C. Ezzell, *Juvenile Court Cases in North Carolina, 1929–1934* (Raleigh, N.C.: State Board of Charities and Public Welfare, 1937), 3, 7, 8, 11–18, quote on page 17. This text is located in the Outer Banks History Center, North Carolina State Archives, Manteo, North Carolina.

6. Ibid., 12–13, 17–20.

7. Ibid., 15–18, quote on page 17.

8. Ibid, 27–30, quote on page 29.

9. Wiley Britton Sanders, *Juvenile Courts in North Carolina* (Chapel Hill: Univ. of North Carolina Press, 1948), 166–171, quote on pages 169–170.

10. J. G. Wooten, "General Order No. 1," by the City of Winston-Salem, March 30, 1934, 5, in "Concerning Problems of Delinquency: Juvenile Detention," Correspondence, 1934–1935, box 86, Institutions and Corrections, State Board of Charities and Public Welfare, North Carolina State Archives, Raleigh, North Carolina.

11. Brice, "The Treatment of African American Girls in Progressive Era North Carolina's Juvenile Justice System," 8–11, 85; Cahn, *Sexual Reckonings*, 96; "Charlotte Hawkins Brown Museum: A Timeline of Achievement," at North Carolina Historic Sites, Division of State Historic Sites and Properties, North Carolina Department of Cultural Resources, www.nchistoricsites.org/chb /civic-life.htm, accessed January 10, 2015.

12. Sanders and Ezzell, *Juvenile Court Cases in North Carolina, 1929–1934*, 13–14, 18, 28, 35.

13. Ibid., 28–31–35; J. G. Wooten, "General Order No. 1," by the City of Winston-Salem, March 30, 1934, 5, in "Concerning Problems of Delinquency: Juvenile Detention," Correspondence, 1934–1935, box 86, Institutions and Corrections, State Board of Charities and Public Welfare, North Carolina State Archives, Raleigh, North Carolina, pages 2–4.

14. Dodge, *"Whores and Thieves of the Worst Kind,"* 4, 8. See also Estelle B. Freedman, *Their Sisters' Keepers: Prison Reform in America, 1830–1930* (Ann Arbor: Univ. of Michigan Press, 1981).

15. Regina Kunzel, *Criminal Intimacy: Prison and the Uneven History of Modern American Sexuality* (Chicago: Univ. of Chicago Press, 2008), 8–9, 51, 53, 58–59, 87, 89; Estelle B. Freedman, "The Prison Lesbian: Race, Class, and the Construction of the Aggressive Female Homosexual, 1915–1965," *Feminist Studies* 22, no. 2 (1996): 398–399. For an intriguing discussion of the "heterosexuality matrix," see Michael A. Rembis, "Beyond the Binary: Rethinking the Social Model of Disabled Sexuality," *Sexual Disability* 28, no. 1 (March 2010): 51–60.

16. Report of Investigation at Samarcand Manor, page 2, and statement of Agnes MacNaughton, state investigation titled "Part II," both in "Samarcand Manor, 1931," box 164, State Board of Charities and Public Welfare, Institutions and Corrections, State Charitable, Penal and Correctional Institutions, Samarcand, State Archives of North Carolina, Raleigh, North Carolina, 76–77; Kunzel, *Criminal Intimacy*, 51–52, 59, 89.

17. Rafter, *Partial Justice*, xxiv–xxv.

18. Mary Belle Harris, *I Knew Them in Prison* (New York: Viking, 1936), 22.

19. Kate Richards O'Hare, *In Prison* (Seattle: Univ. of Washington Press, 1923), ix, 55, 62–67, quote on page 72.

20. Anne M. Butler, *Gendered Justice in the American West: Women Prisoners in Men's Penitentiaries* (Urbana: Univ. of Illinois Press, 1997), 6, 26–31; Dodge, *"Whores and Thieves of the Worst Kind,"* 3, 4, 15, quote on page 14; Mark Colvin, *Penitentiaries, Reformatories, and Chain Gangs: Social Theory and the History of Punishment in Nineteenth-Century America* (New York: St. Martin's, 1997).

21. Dodge, *"Whores and Thieves of the Worst Kind,"* 16, quotes on pages 17, 117. See also Cesare Lombroso and William Ferrero, *The Female Offender* (1895; reprint, New York: Appleton, 1900); Marcus Kavanaugh, *The Criminal and His Allies* (Indianapolis: Bobbs-Merrill, 1928).

22. *Biennial Report of the Superintendent of the State Prison Department, June 30, 1932* (Raleigh, N.C.: N.p., 1932), 26, 67–70, 85, www.archive.org/stream/biennialreport193032nort#page/n7/mode/2up, accessed June 28, 2013.

23. Dan C. Boney to George Ross Pou, September 16, 1931, George Ross Pou to Dan C. Boney, September 19, 1931, and George Ross Pou to O. Max Gardner, September 19, 1931, in "New Prison," 1930–1932, General Correspondence, box 96, Governor's Papers: O. Max Gardner, 1929–1933, North Carolina State Archives, Raleigh, North Carolina.

24. Ibid.

25. F. Lovell Bixby to Capus M. Waynick, October 15, 1935, Capus M. Waynick to J.C.B. Ehringhaus, October 21, 1935, and J.C.B. Ehringhaus to Capus Waynick, October 23, 1935, all in "Prisoners," 1933–1936, Prison Department, Records of the State Highway and Public Works Commission, box 155, General Correspondence, Governor's Papers: J.C.B. Ehringhaus, North Carolina State Archives, Raleigh, North Carolina; "State Use of Prison Made Goods: A Report to the Conference of Governors From National Committee on Prisons and Prison Labor," June 1–3, 1931, page 1, in "Prisoners," 1929–1932, General Correspondence, Governor's Papers: O. Max Gardner, State Archives of North Carolina, Raleigh, North Carolina; O. Max Gardner to C. H. Wheatley, December 5, 1929, C. R. Wheatley to George Ross Pou, November 27, 1929, Ira Plemmons to O. Max Gardner, November 17, 1932, in "Prison Labor," 1929–1932, General Correspondence, box 96, Governor's Papers: O. Max Gardner, 1929–1933, State Archives of North Carolina, Raleigh, North Carolina. Capus M. Waynick to William H. Carroll, September 13, 1935, Thomas G. Neal to J.C.B. Ehringhaus, Western Union Telegram, September 16, 1935; J.C.B. Ehringhaus to Thomas G. Neal, Western Union Telegram, September 17, 1935, all in "Women's Prison, 1935," Prison Department, Records of the State Highway and Public Works Commission, box 155, Governor's Papers, J.C.B. Ehringhaus, State Archives of North Carolina, Raleigh, North Carolina.

26. *Biennial Report of the Superintendent of the State Prison Department, June 30, 1932*, 50–51, www.archive.org/stream/biennialreport193032nort#page/n7/mode/2up, accessed June 28, 2013; "North Carolina considers compensating victims of its eugenics programs," August 4, 2011, NewsyType.com, www.newsytype.com/9768-north-carolina-eugenics/, accessed June 28, 2013.

27. Descriptive Register of Prisoners, Central (State) Prison, January 8, 1929–May 28, 1934, vol. 13, Prison Department, Department of Corrections, Records of the State Highway and Public Works Commission, State Archives of North Carolina, Raleigh, North Carolina; Cases 25502–25513, Commutation Book, vol. 17, Assistant Director in Charge of Rehabilitation, Prison Department, Department of Corrections, Records of the State Highway and Public Works Commission, State Archives of North Carolina, Raleigh, North Carolina.

28. R. F. Beasley, "Grand Jury Investigation Samarcand," and "An Open Letter to Grand Jury Foremen," n.d., newspaper clippings in miscellaneous scrapbook, Samarcand Manor Scrapbook Collection, Eagle Springs, North Carolina; George Herr to Armstrong, circa February 4, 1943, George Herr to Armstrong, February 22, 1943, in miscellaneous scrapbook, Samarcand Manor

Scrapbook Collection, Eagle Springs, North Carolina; "Investigation Samarcand Ordered," *Moore County (N.C.) News,* January 21, 1943.

29. "Investigation Samarcand Ordered," *Moore County (N.C.) News,* January 21, 1943; Frances Newsom, "Samarcand Clothes Is $4 per Girl per Year," *Winston-Salem (N.C.) Journal,* February 1, 1943.

30. "Report of the Special Committee of the Moore County Grand Jury of Its Investigation of Samarcand Manor, Eagle Springs," May 1943, in miscellaneous scrapbook, Samarcand Manor Scrapbook Collection, Eagle Springs, North Carolina, 2, 4, 6.

31. Ibid., 4, 6, 8.

EPILOGUE

1. Ann Brock, "Samarkand to Become Public Safety Training Center," *Sandhills Tribune,* Carthage, North Carolina, August 1, 2013, www.sandhillstribune.com/index.php/news/news-2 /item/2857-samarkand-to-become-public-safety-training-center/2857-samarkand-to-become -public-safety-training-center, accessed April 26, 2014.

2. H. M. DuBose Jr. to J.C.B. Ehringhaus, July 6, 1934, and the deposition of James Cooper, July 6, 1934, and R. Eugene Brown, "Report on Progress Made in Reorganization of Jackson Training School," December 21, 1934, all in "Stonewall Jackson Training School, Concord, Correspondence, 1933–1935," Governor's Papers, J.C.B. Ehringhaus, box 10, State Archives of North Carolina, Raleigh, North Carolina.

3. Elizabeth Leland, "Stonewall Jackson Secrets: 'Children Against Monsters,'" *Charlotte (N.C.) Observer,* October 5, 2013.

4. Ibid.; James R. Tompkins, "Hell Without Fire: A Report on Children" (Boone, N.C.: Commissioned by the North Carolina Governor's Advocacy Commission for Children and Youth, January 1975), copy provided to the author.

5. Ibid.

6. John Krahnert, "Report Alleges Sex Acts at Samarkand," *The Pilot,* Southern Pines, North Carolina, January 10, 2010, High Point, North Carolina; "Sex Abuse High at Youth Detention Centers," CBS News, January 7, 2010, www.cbsnews.com/news/sex-abuse-high-at-youth-detention -centers, accessed April 26, 2014.

7. Michael Biesecker, "Juvenile Inmates Report Sex Acts: State Disputes Study's Findings," *Raleigh (N.C.) News and Observer,* January 9, 2010, www.newsobserver.com/2010/01/09/275466 /juvenile-inmates-report-sex-acts.html, accessed April 26, 2014; North Carolina Department of Juvenile Justice and Delinquency Prevention, "Secretary Linda Hayes' Statement on the 'Sexual Victimization in Juvenile Facilities Reported by Youth' Report Release," January 9, 2010.

8. Tim Wu, "Fifty-Five Bodies and Zero Trials at the Florida School for Boys," *New Yorker,* January 30, 2014: http://www.newyorker.com/online/blogs/newsdesk/2014/01/prosecutors-are-failing-the-victims-of-floridas-notorious-reform-school.html, accessed April 26, 2014; Annie E. Casey Foundation, "No Place for Kids: The Case for Reducing Juvenile Incarceration, Issue Brief" (Baltimore, Md.: Annie E. Casey Foundation, 2011), http://www.aecf.org/resources/no-place-for -kids-full-report, accessed May 11, 2015.

9. Mark W. Lipsey et al., "Improving the Effectiveness of Juvenile Justice Programs: A New Perspective on Evidence-Based Practice," Center for Juvenile Justice Reform, Georgetown Pub-

lic Policy Institute, Georgetown University, December 2010, 5–6, cjjr.georgetown.edu/pdfs/ebp/ ebppaper.pdf, accessed April 14, 2014; Giudi Weiss, "The Fourth Wave: Juvenile Justice Reforms for the Twenty-First Century," Commissioned by the National Campaign to Reform State Juvenile Justice Systems for the Juvenile Justice Funders' Collaborative, Winter 2013: www.modelsfor change.net/publications/530, accessed May 11, 2015.

10. Lipsey et al., "Improving the Effectiveness of Juvenile Justice Programs," 8–11; Jeffrey A. Butts and Douglas N. Evans, "Resolution, Reinvestment, and Realignment: Three Strategies for Changing Juvenile Justice" (New York, N.Y.: Research and Evaluation Center, John Jay College of Criminal Justice, City University of New York, 2011), ii, 1, www.johnjayresearch.org/wp-content /uploads/2011/09/rec20111.pdf, accessed April 14, 2014; "Legislation," Office of Juvenile Justice and Delinquency Prevention, www.ojjdp.gov/about/legislation.html, accessed April 14, 2014.

11. Fiscal Research Division, Fiscal Brief, "Justice and Public Safety Subcommittee: FY 2012–13 Budget Highlights," North Carolina General Assembly House and Senate Appropriations Subcommittees on Justice and Public Safety, October 2012, www.ncleg.net/fiscalresearch /fiscal_briefs/2012_Session_Budget_Fiscal_Briefs/JPS_2012_Fiscal_Brief_2012-08-08.pdf, accessed April 14, 2014.

12. Ibid.; Lynn Bonner et al., "Budget Takes Shape in House Committees," *Raleigh (N.C.) News and Observer,* May 25, 2012: www.newsobserver.com/2012/05/25/2087594/house-endorses -budget-plan-that.html, accessed April 11, 2014; Thomasi McDonald, "Reforms Credited for Driving Juvenile Crime Down in North Carolina," *Raleigh (N.C.) News and Observer,* October 7, 2012.

13. Buddy Howell, "Misguided Juvenile Cuts," *Raleigh (N.C.) News and Observer,* January 4, 2013, as reported on the International Child and Youth Care Network, http://www.cyc-net.org /opinion/opinion-130107.html, accessed on May 11, 2015; "Adult-Juvenile Merger Concerns Advocates," September 2, 2013; Linda Hayes, "Wrong Move on Juvenile Justice: Eliminating a Division Focused Solely on Young Offenders Will Cost More in the Long Run," September 22, 2013.

BIBLIOGRAPHY

MANUSCRIPT COLLECTIONS

David M. Rubenstein Rare Book and Manuscript Collection. Duke University. Durham, North Carolina.
  G. Hope Summerell Chamberlain Papers.
East Carolina Manuscript Collection. Joyner Library. East Carolina University Greenville, North Carolina.
  Kate Ancrum Burr Johnson Papers.
North Carolina State Archives. Outer Banks History Center. Manteo, North Carolina.
  The David Stick Library
North Carolina State Archives. Raleigh, North Carolina.
  Governor J.C.B. Ehringhaus Papers, 1933–1937.
  Governor O. Max Gardner Papers, 1929–1933.
  Nell Battle Lewis Papers.
  Records of the Department of Public Safety.
  Records of the Eugenics Commission.
  Records of the North Carolina General Assembly, Session 1917, 2014.
  Records of the State Board of Health.
  Records of the State Board of Charities and Public Welfare.
  Records of the State Charitable, Penal, and Correctional Institutions.
  Records of the State Department of Juvenile Justice and Delinquency Programs.
  Records of the State Highway and Public Works Commission.
  Records of Samarcand Manor. Division of Adult Correction and Juvenile Justice, Department of Public Safety, Samarcand Manor School, Eagle Springs, North Carolina.
  Samarcand Manor Scrapbook Collection.
  Uncatalogued Files and Folders.

GOVERNMENT DOCUMENTS

"Legislation." Office of Juvenile Justice and Delinquency Prevention. www.ojjdp.gov /about/legislation.html. Accessed April 14, 2014.

Loving V. Virginia, 388 U.S. 1.

Manual for the Various Agents of the United States Interdepartmental Social Hygiene Board. Washington, D.C.: GPO, 1920.

National Climatic Data Center. Climatic Data Online. www.ncdc.noaa.gov/cdo-web/. Accessed November 1, 2010.

The North Carolina Code of 1939. Charlottesville, Va.: Michie Company, Law Publishers, 1939.

North Carolina State Board of Charities and Public Welfare, *Biennial Report of the North Carolina State Board of Charities and Public Welfare, 1920–1922.* Raleigh, N.C.: The Board, 1922.

North Carolina State Board of Charities and Public Welfare, *Biennial Report of the North Carolina State Board of Charities and Public Welfare, 1922–1924.* Raleigh, N.C.: The Board, 1924.

North Carolina State Board of Charities and Public Welfare, *Biennial Report of the North Carolina State Board of Charities and Public Welfare, 1926–1928.* Raleigh, N.C.: Mitchell Printing Co., 1928.

North Carolina State Board of Charities and Public Welfare, *Biennial Report of the Superintendent of the State Prison Department, June 30, 1932.* Raleigh, N.C.: N.d., 1932. www.archive.org/stream/biennialreport193032nort#page/n7/mode/2up. Accessed June 28, 2013.

Public Laws and Resolutions of the State of North Carolina, passed by the General Assembly at its Session of 1911. Raleigh, N.C.: Uzzell, 1911.

Public Laws and Resolutions of the State of North Carolina, passed by the General Assembly at its Session of 1913. Raleigh, N.C.: Edwards and Broughton, 1913.

Public Laws and Resolutions of the State of North Carolina, passed by the General Assembly at its Session of 1917. Raleigh, N.C.: Edwards and Broughton, 1917.

Public Laws and Resolutions of the State of North Carolina, passed by the General Assembly at its Session of 1919. Raleigh, N.C.: Uzzell, 1919.

Public Laws and Resolutions of the State of North Carolina, passed by the General Assembly at its Session of 1921. Raleigh, N.C.: Mitchell Printing Co. 1921.

Public Laws and Resolutions of the State of North Carolina, passed by the General Assembly at its Session of 1923. Raleigh, N.C.: State Printer, 1923.

Public Laws and Resolutions of the State of North Carolina, passed by the General Assembly at its Session of 1929. Fort Wayne, Ind.: Fort Wayne Printing Co., 1929.

Public Laws and Resolutions of the State of North Carolina, passed by the General Assembly at its Session of 1931. Charlotte, N.C.: Observer Printing House, 1931.

Public Laws and Resolutions of the State of North Carolina, passed by the General Assembly at its Session of 1933. Charlotte, N.C.: Observer Printing House, 1933.

Sanders, Wiley Britton, and William C. Ezzell. *Juvenile Court Cases in North Carolina, 1929–1934.* Raleigh, N.C.: State Board of Charities and Public Welfare, 1937.

A Square Deal for the Child: Juvenile Courts in North Carolina. Raleigh, N.C.: North Carolina State Board of Charities and Public Welfare, 1923.

State Laws Governing Public Welfare Work in North Carolina. Raleigh, N.C.: North Carolina State Board of Charities and Public Welfare, 1923.

Thompson, Drury B. *Juvenile Courts in North Carolina.* Raleigh, N.C.: State Board of Public Welfare, 1951.

NEWSPAPERS

Chapel Hill Weekly. Chapel Hill, North Carolina.

Charlotte News. Charlotte, North Carolina.

Charlotte Observer. Charlotte, North Carolina.

Charlotte Sunday Observer. Charlotte, North Carolina.

Fayetteville Observer. Fayetteville, North Carolina.

Gastonia Daily Gazette. Gastonia, North Carolina.

Los Angeles Times. Los Angeles, California.

Moore County News. Carthage, North Carolina.

News and Observer. Raleigh, North Carolina.

New York Times. New York, New York.

Orphans' Friend and Masonic Journal. Oxford, North Carolina.

The Pilot. Southern Pines, North Carolina.

Pinehurst Outlook. Pinehurst, North Carolina.

Rocky Mount Telegram. Rocky Mount, North Carolina.

Sandhills Tribune. Carthage, North Carolina.

Winston-Salem Journal. Winston-Salem, North Carolina.

WORLD WIDE WEB

Butts, Jeffrey A., and Douglas N. Evans. "Resolution, Reinvestment, and Realignment: Three Strategies for Changing Juvenile Justice." New York, N.Y.: Research and Evaluation Center, John Jay College of Criminal Justice, City University of New York, 2011. www.johnjayresearch.org/wp-content/uploads/2011/09/rec20111.pdf. Accessed April 14, 2014.

"Charlotte Hawkins Brown Museum: A Timeline of Achievement." North Carolina Historic Sites. North Carolina Department of Cultural Resources. www.nchistoric sites.org/chb/time-line.htm. Accessed January 10, 2015.

Hutchison, Courtney. "Sterilizing the Sick, Poor to Cut Welfare Costs: North Carolina's History of Eugenics." August 4, 2011. Abcnews.go.com. Accessed December 18, 2013.

"North Carolina Considers Compensating Victims of Its Eugenics Programs." August 2, 2011. NewsyType.com. www.newsytype.com/9768-north-carolina-eugenics. Accessed June 28, 2013.

"Sex Abuse High at Youth Detention Centers." CBS News. January 7, 2010. www.cbsnews .com/news/sex-abuse-high-at-youth-detention-centers. Accessed April 26, 2014.

Welty, Jeff. "The Death Penalty in North Carolina: History and Overview." April 2012. www.unc.edu/~fbaum/teaching/articles/Welty-DP-overview.pdf. Accessed January 2, 2014.

BOOKS, ARTICLES, AND DISSERTATIONS

Alexander, Ruth M. *The "Girl Problem": Female Sexual Delinquency in New York, 1900–1930.* Ithaca: Cornell Univ. Press, 1998.

American Social Hygiene Association. "The American Social Hygiene Association, Inc. [1918]." *Journal of Social Hygiene* 4 (1918): 138a.

Annie E. Casey Foundation, "No Place for Kids: The Case for Reducing Juvenile Incarceration, Issue Brief." Baltimore, Md.: Annie E. Casey Foundation, 2011. http:// www.aecf.org/resources/no-place-for-kids-full-report. Accessed May 11, 2015.

Beck, E. M., James L. Massey, and Stewart E. Tolnay. "The Gallows, the Mob, and the Vote: Lethal Sanctioning of Blacks in North Carolina and Georgia, 1882 to 1930." *Law and Society Review* 23, no. 2 (1989): 317–331.

Bederman, Gail. *Manliness and Civilization: A Cultural History of Gender and Race in the United States, 1880–1917.* Chicago: Univ. of Chicago Press, 1996.

Begos, Kevin, et al. *Against Their Will: North Carolina's Sterilization Program.* Apalachicola, Fla.: Gray Oak Books, 2012.

Belknap, Joanne. *The Invisible Woman: Gender, Crime, and Justice.* Belmont, Calif.: Wadsworth, 1996.

Bickford, Annette Louise. "Imperial Modernity, National Identity and Capital Punishment in the Samarcand Arson Case, 1931." *Nations and Nationalism* 13, no. 3 (July 2007): 437–460.

"Birdie Dunn." *American Journal of Nursing* 46, no. 9 (September 1946): 640.

Black, Edwin. *War Against the Weak: Eugenics and America's Campaign to Create a Master*

Race. New York: Four Walls Eight Windows, 2003.

Brandt, Allan M. *No Magic Bullet: A Social History of Venereal Disease in the United States Since 1880.* New York: Oxford Univ. Press, 1987.

Brice, Tanya Smith. "The Treatment of African American Girls in Progressive Era North Carolina's Juvenile Justice System, 1890–1930." Ph.D. diss., School of Social Work, University of North Carolina at Chapel Hill, 2003.

Bristow, Nancy K. *Making Men Moral: Social Engineering During the Great War.* New York: New York Univ. Press, 1996.

Brown, Kathleen M. *Good Wives, Nasty Wenches, and Anxious Patriarchs.* Chapel Hill: Univ. of North Carolina Press, 1996.

Brundage, W. Fitzhugh. *Lynching in the New South: Georgia and Virginia, 1880–1930.* Urbana: Univ. of Illinois Press, 1993.

Butler, Anne M. *Gendered Justice in the American West: Women Prisoners in Men's Penitentiaries.* Urbana: Univ. of Illinois Press, 1997.

Byerly, Victoria, and Cletus E. Daniel. *Hard Times Cotton Mill Girls: Personal Histories of Womanhood and Poverty in the South.* Ithaca: Cornell Univ. Press, 1986.

Bynum, Victoria E. *Unruly Women: The Politics of Social and Sexual Control in the Old South.* Chapel Hill: Univ. of North Carolina Press, 1992.

Cahn, Susan K. *Sexual Reckonings: Southern Girls in a Troubling Age.* Boston: Harvard Univ. Press, 2007.

——. "Spirited Youth or Fiends Incarnate: The Samarcand Arson Case and Female Adolescence in the American South." *Journal of Women's History* 9, no. 4 (winter 1998): 152–180.

Caldwell, Erskine. *God's Little Acre.* New York: Viking, 1933.

——. *Tobacco Road.* Cutchogue, N.Y.: Buccaneer Books, 1932.

Carlton-LaNey, I. B., ed. *African American Leadership: An Empowerment Tradition in Social Welfare History.* Washington, D.C.: NASW Press, 2001.

Castles, Katherine. "Quiet Eugenics: Sterilization in North Carolina's Institutions for the Mentally Retarded, 1945–1965." *Journal of Southern History* 68, no. 4 (November 2002): 849–878.

Chamberlain, Hope Summerell. *This Was Home.* Chapel Hill: Univ. of North Carolina Press, 1938.

Clarke, James W. "Without Fear or Shame: Lynching, Capital Punishment and the Subculture of Violence in the American South." *British Journal of Political Science* 28, no. 2 (April 1998): 269–289.

Clatton [*sic*, Clayton], S. Lillian. "The Social Training of the Private Duty Nurse." *American Journal of Nursing* 17, no. 6 (March 1917): 506–508.

Clement, Elizabeth Alice. *Love for Sale: Courting, Treating, and Prostitution in New York City, 1900–1945.* Chapel Hill: Univ. of North Carolina Press, 2006.

Colvin, Mark. *Penitentiaries, Reformatories, and Chain Gangs: Social Theory and the History of Punishment in Nineteenth-Century America.* New York: St. Martin's, 1997.

Connor, R.D.W., ed. *North Carolina Manual.* Raleigh, N.C.: Edwards and Broughton, 1917.

Cooper, John Milton, Jr., and Charles E. Neu. *The Wilson Era: Essays in Honor of Arthur S. Link.* Arlington Heights, Ill.: Harlan Davidson, 1991.

Cott, Nancy F. *The Grounding of Modern Feminism.* New Haven: Yale Univ. Press, 1987.

Cotten, Sallie Southall. *History of the North Carolina Federation of Women's Clubs, 1901–1925.* Raleigh, N.C.: Edwards and Broughton, 1925.

Cox, Annette. "The Loray, North Carolina's 'Million Dollar Mill': The 'Monstrous Hen' of Southern Textiles." *North Carolina Historical Review* 89, no. 3 (July 2012): 241–275.

Crane, Stephen. *Maggie: A Girl of the Streets, and Other New York Writings.* 1893. Reprint, New York: Modern Library, 2001.

Currell, Susan, and Christina Cogdell, eds. *Popular Eugenics: National Efficiency and American Mass Culture in the 1930s.* Athens: Ohio Univ. Press, 2006.

DeLuzio, Crista. *Female Adolescence in American Scientific Thought, 1830–1930.* Baltimore: Johns Hopkins Univ. Press, 2007.

Dodge, L. Mara. *"Whores and Thieves of the Worst Kind": A Study of Women, Crime, and Prisons, 1835–2000.* DeKalb: Northern Illinois Univ. Press, 2006.

Edwards, Laura F. *Scarlett Doesn't Live Here Anymore: Southern Women in the Civil War Era.* Urbana: Univ. of Illinois Press, 2000.

Escott, Paul D., ed. *North Carolinians in the Era of the Civil War and Reconstruction.* Chapel Hill: Univ. of North Carolina Press, 2008.

Evans, Sara M. *Born for Liberty: A History of Women in America.* New York: Free Press, 1989.

Falconer, Martha P. "Causes of Delinquency Among Girls." *Annals of the American Academy of Political and Social Science* 36, no. 1 (July 1910): 77–79.

———. "Industrial Schools for Girls and Women." *Social Hygiene* 3, no. 3 (July 1917): 323–330.

———. "The Part of the Reformatory Institution in the Elimination of Prostitution." *Social Hygiene* 5, no. 1 (January 1919): 1–9.

———. "Report of the Committee on Social Hygiene of the National Conference of Charities and Correction." *Social Hygiene* 1 (September 1915): 520–521.

Fass, Paula S. *The Damned and the Beautiful: American Youth in the 1920's.* New York: Oxford Univ. Press, 1977.

Fitzgerald, F. Scott. *Flappers and Philosophers.* New York: Scribner and Sons, 1920.

Freedman, Estelle B. "The Prison Lesbian: Race, Class, and the Construction of the Aggressive Female Homosexual, 1915–1965." *Feminist Studies* 22, no. 2 (1996): 397–423.

———. *Their Sisters' Keepers: Prison Reform in America, 1830–1930.* Ann Arbor: Univ. of Michigan Press, 1981.

Giddings, Paula J. *When and Where I Enter: The Impact of Black Women on Race and Sex in America.* New York: William Morrow Paperbacks, 2007.

Gilmore, Glenda Elizabeth. *Gender and Jim Crow: Women and the Politics of White Supremacy in North Carolina, 1896–1920.* Chapel Hill: Univ. of North Carolina Press, 1996.

Ginzberg, Lori D. *Women and the Work of Benevolence: Morality, Politics, and Class in the Nineteenth-Century United States.* New Haven: Yale Univ. Press, 1992.

Gourley, Catherine. *Flappers and the New American Woman.* Minneapolis: Twenty-First Century Books, 2008.

Green, Elna C. *Southern Strategies: Southern Women and the Woman Suffrage Question.* Chapel Hill: Univ. of North Carolina Press, 1997.

Hall, G. Stanley. *Adolescence: Its Psychology and Its Relation to Physiology, Anthropology, Sociology, Sex, Crime, Religion and Education.* 2 vols. New York: Appleton, 1904.

Hall, Jacquelyn Dowd. "Disorderly Women: Gender and Labor Militancy in the Appalachian South." *Journal of American History* 73, no. 2 (September 1986): 354–382.

———. *Revolt Against Chivalry: Jessie Daniel Ames and the Women's Campaign Against Lynching.* New York: Columbia Univ. Press, 1993.

Hall, Jacquelyn Dowd, et al. *Like a Family: The Making of a Southern Cotton Mill World.* Chapel Hill: Univ. of North Carolina Press, 2000.

Harris, Mary Belle. *I Knew Them in Prison.* New York: Viking, 1936.

Holloway, Pippa. *Sexuality, Politics, and Social Control in Virginia, 1920–1945.* Chapel Hill: Univ. of North Carolina Press, 2006,

Hooks, Bell. *Ain't I a Woman: Black Women and Feminism.* 1981. Cambridge, Mass.: South End Press, 1999.

Johnson, Kate Burr. "Problems of Delinquency Among Girls." *Journal of Social Hygiene* 12, no. 7 (October 1926): 385–397.

Kaufman-Osborn, Timothy V. "Symposium on Gender and the Death Penalty." *Signs: Journal of Women in Culture and Society* 24, no. 4 (summer 1999): 1097–1102.

Kavanaugh, Marcus. *The Criminal and His Allies.* Indianapolis: Bobbs-Merrill, 1928.

Keire, Mara Laura. *For Business and Pleasure: Red-Light Districts and the Regulation of Vice in the United States, 1890–1933.* Baltimore: Johns Hopkins Univ. Press, 2010.

Kevles, Daniel J. *In the Name of Eugenics: Genetics and the Uses of Human Heredity.* Cambridge: Harvard Univ. Press, 1998.

Kline, Wendy. *Building a Better Race: Gender, Sexuality, and Eugenics from the Turn of the Century to the Baby Boom.* Berkeley: Univ. of California Press, 2005.

Krome-Lukens, Anna. "A Great Blessing to Defective Humanity: Women and the Eugenics Movement in North Carolina, 1910–1940." Master's thesis, University of North Carolina at Chapel Hill, 2009.

————. "The Reform Imagination: Gender, Eugenics, and the Welfare State in North Carolina, 1900–1940." Ph.D. diss., University of North Carolina at Chapel Hill, 2014.

Kuhl, Stefan. *The Nazi Connection: Eugenics, American Racism, and German National Socialism.* New York: Oxford Univ. Press, 1994.

Kunzel, Regina G. *Criminal Intimacy: Prison and the Uneven History of Modern American Sexuality.* Chicago: Univ. of Chicago Press, 2008.

————. *Fallen Women, Problem Girls: Unmarried Mothers and the Professionalization of Social Work, 1890–1940.* New Haven: Yale Univ. Press, 1995.

Ladu, Lena B., and K. C. Garrison. "A Study of Emotional Instability and Intelligence of Women in the Penal Institutions of North Carolina." *Social Forces* 10, no. 2 (December 1931): 209–216.

Lallier, Rebecca Ragsdale. "'A Place of Beginning Again': The North Carolina Industrial Farm Colony for Women, 1929–1947." Master's thesis, University of North Carolina at Chapel Hill, 1990.

Larson, Edward J. *Sex, Race, and Science: Eugenics in the Deep South.* Baltimore: John Hopkins Univ. Press, 1996.

Latham, Angela J. *Posing a Threat: Flappers, Chorus Girls, and Other Brazen Performers of the American 1920s.* Hanover, N.H.: Wesleyan Univ. Press, 2000.

Leidholt, Alexander. *Battling Nell: The Life of Southern Journalist Nell Battle Lewis.* Baton Rouge: Louisiana State Univ. Press, 2009.

Lewis, Nell Battle, ed. *Capital Punishment in North Carolina: Special Bulletin Number 10.* Raleigh: North Carolina State Board of Charities and Public Welfare, 1929.

Lichtenberger, J. P., ed. *The Annals: War Relief Work.* Philadelphia: American Academy of Political and Social Science, 1918.

Link, William A. *The Paradox of Southern Progressivism, 1880–1930.* Chapel Hill: Univ. of North Carolina Press, 1992.

Lipsey, Mark W., et al. "Improving the Effectiveness of Juvenile Justice Programs: A New Perspective on Evidence-Based Practice." Center for Juvenile Justice Reform, Georgetown Public Policy Institute, Georgetown University, December 2010. cjjr. georgetown.edu/pdfs/ebp/ebppaper.pdf. Accessed April 14, 2014.

Lombroso, Cesare, and William Ferrero. *The Female Offender.* 1895. Reprint, New York: Appleton, 1900.

Luker, Kristin. "Sex, Social Hygiene, and the State: The Double-Edged Sword of Social Reform." *Theory and Society* 27, no. 5 (October 1998): 601–634.

Mangan J. A., and Roberta J. Park, eds. *From "Fair Sex" to Feminism: Sport and the Socialization of Women in the Industrial and Post-Industrial Eras.* London: Cass, 1987.

McLaurin, Melton, and Anne Russell. *The Wayward Girls of Samarcand: A True Story of the American South.* Wilmington, N.C.: Bradley Creek Press, 2012.

McRae, Elizabeth Gillespie. "To Save a Home: Nell Battle Lewis and the Rise of Southern Conservatism." *North Carolina Historical Review* 81, no. 3 (July 2004): 261–287.

Nathanson, Constance. *Dangerous Passage: The Social Control of Sexuality in Women's Adolescence*. Philadelphia: Temple Univ. Press, 1991.

Noll, Steven. *Feeble-Minded in Our Midst: Institutions for the Mentally Retarded in the South, 1900–1940*. Chapel Hill: Univ. of North Carolina Press, 1995.

"Nursing News and Announcements." *American Journal of Nursing* 12, no. 9 (June 1912): 752.

Odem, Mary E. *Delinquent Daughters: Protecting and Policing Adolescent Female Sexuality in the United States, 1885–1920*. Chapel Hill: Univ. of North Carolina Press, 1995.

O'Hare, Kate Richards. *In Prison*. Seattle: Univ. of Washington Press, 1923.

Ogburn, William Fielding. Review of *Applied Eugenics* by Paul Popenoe and Roswell Hill Johnson. *Political Science Quarterly* (September 1921): 533–534.

———. *Social Change with Respect to Culture and Original Nature*. New York: Viking, 1928.

Ordover, Nancy. *American Eugenics: Race, Queer Anatomy, and the Science of Nationalism*. Minneapolis: Univ. of Minnesota Press, 2003.

O'Shea, Kathleen. *Women and the Death Penalty in the United States, 1900–1998*. Westport, Conn.: Praeger, 1999.

Peiss, Kathy. *Cheap Amusements: Working Women and Leisure in Turn-of-the-Century New York*. Philadelphia: Temple Univ. Press, 1986.

Phillips, Charles D. "Exploring Relations among Forms of Social Control: The Lynching and Execution of Blacks in North Carolina, 1889–1918." *Law and Society Review* 21, no. 3 (1987): 361–374.

———. "Social Structure and Social Control: Modeling the Discriminatory Execution of Blacks in Georgia and North Carolina, 1925–1935." *Social Forces* 65, no. 2 (December 1986): 458–475.

Proceedings of the Quarterly Conference: New Jersey State Board of Control of Institutions and Agencies, 1922. Memphis: General Books, 2010.

Rafter, Nicole Hahn. *Partial Justice: Women in State Prisons, 1800–1935*. Boston: Northeastern Univ. Press, 1985.

Rauschenbusch, Walter. *Christianity and the Social Crisis*. Library of Theological Ethics. Louisville, Ky.: Westminster John Knox Press, 1992.

Rembis, Michael A. "Beyond the Binary: Rethinking the Social Model of Disabled Sexuality." *Sexual Disability* 28, no. 1 (March 2010): 51–60.

———. *Defining Deviance: Sex, Science, and Delinquent Girls, 1890–1960*. Urbana: Univ. of Illinois Press, 2013.

———. "'I Ain't Been Reading While on Parole': Experts, Mental Tests, and Eugenic Commitment Law in Illinois, 1890–1940." *History of Psychology* 7, no. 3 (August 2004): 225–247.

Robson, Grace M. "Samarcand Manor," in "First Juvenile Delinquent Training School in North Carolina Was Opened in 1909," *Public Welfare News* 7, no. 1 (March 1944): 4–5, 23–30.

Rowson, Susanna. *Charlotte Temple.* Edited by Cathy N. Davidson. Oxford: Oxford Univ. Press, 1986.

Salmond, John A. "The Burlington Dynamite Plot: The 1934 Textile Strike and Its Aftermath in Burlington, North Carolina." *North Carolina Historical Review* 75, no. 4 (October 1998): 398–434.

———. *Gastonia, 1929: The Story of the Loray Mill Strike.* Chapel Hill: Univ. of North Carolina Press, 1995.

———. *The General Textile Strike of 1934: From Maine to Alabama.* Columbia: Univ. of Missouri Press, 2002.

Sanders, Wiley Britton. *Juvenile Courts in North Carolina.* Chapel Hill: Univ. of North Carolina Press, 1948.

Scheiner, Georganne. *Signifying Female Adolescence: Film Representations and Fans, 1920–1950.* Westport, Conn.: Praeger, 2000.

Schoen, Johanna. *Choice and Coercion: Birth Control, Sterilization, and Abortion in Public Health and Welfare.* Chapel Hill: Univ. of North Carolina Press, 2005.

———. "From the Footnotes to the Headlines: Sterilization Apologies and Their Lessons." *Sexuality Research and Social Policy* 3, no. 3 (September 2006): 7–22.

Scott, Anne Firor. *The Southern Lady: From Pedestal to Politics, 1830–1930.* 1970. Charlottesville: Univ. of Virginia Press, 1995.

Scott, Joan Wallach. "Gender: A Useful Category of Historical Analysis." *American Historical Review* 91, no. 5 (December 1986): 1053–1075.

Seibel, Emily. "Nell Battle Lewis in the Progressive Era: Prison and Capital Punishment Reform." Master's thesis, University of North Carolina at Asheville, 2009.

Sims, Anastatia. *The Power of Femininity in the New South: Women's Organizations and Politics in North Carolina, 1880–1930.* Columbia: Univ. of South Carolina Press, 1997.

Snitow, Ann, Christine Stansell, and Sharon Thompson, eds. *Powers of Desire: The Politics of Sexuality.* New York: Monthly Review, 1983.

Solinger, Rickie. "Layering the Lenses: Toward Understanding Reproductive Politics in the United States." *Journal of Women's History* 25 (winter 2013): 101–112.

Stern, Alexandra. *Eugenic Nation: Faults and Frontiers of Better Breeding in Modern America.* Berkeley: Univ. of California Press, 2005.

Streib, Victor. "Death Penalty for Female Offenders." *University of Cincinnati Law Review* 58 (1990): 845–880.

Tompkins, James R. "Hell Without Fire: A Report on Children." Boone, N.C.: Commissioned by the North Carolina Governor's Advocacy Commission for Children and Youth, January 1975. Copy provided to the author.

Vandiver, Margaret. *Lethal Punishment: Lynchings and Legal Executions in the South.* New Brunswick, N.J.: Rutgers Univ. Press, 2005.

Walkowitz, Judith R. *City of Dreadful Delight: Narratives of Sexual Danger in Late-Victorian London.* Chicago: Univ. of Chicago Press, 1992.

Weiss, Giudi. "The Fourth Wave: Juvenile Justice Reforms for the Twenty-First Century." Commissioned by the National Campaign to Reform State Juvenile Justice Systems for the Juvenile Justice Funders' Collaborative. Winter 2013. www.models forchange.net/publications/530.

Welch, Paula D. *History of American Physical Education and Sport,* 3rd ed. Springfield, Ill.: Thomas, 2004.

Wertheimer, John, Brian Luskey, et al. "'Escape of the Match-Strikers': Disorderly North Carolina Women, the Legal System, and the Samarcand Arson Case of 1931." *North Carolina Historical Review* 75, no. 4 (October 1998): 435–460.

Whalen, Robert Weldon. *"Like Fire in Broom Straw": Southern Journalism and the Textile Strikes of 1929–1931.* Westport, Conn.: Greenwood, 2001.

———. "Recollecting the Cotton Mill Wars: Proletarian Literature of the 1929–1931 Southern Textile Strikes." *North Carolina Historical Review* 75, no. 4 (October 1998): 370–397.

Wheeler, Marjorie Spruill. *New Women of the New South: The Leaders of the Woman Suffrage Movement in the Southern States.* New York: Oxford Univ. Press, 1993.

Woodside, Moya. *Sterilization in North Carolina: A Sociological and Psychological Study.* Chapel Hill: Univ. of North Carolina Press, 1950.

Wray, Matt. *Not Quite White: White Trash and the Boundaries of Whiteness.* Durham, N.C.: Duke Univ. Press, 2006.

Wu, Tim. "Fifty-Five Bodies and Zero Trials at the Florida School for Boys," *New Yorker,* January 30, 2014. www.newyorker.com/online/blogs/newsdesk/2014/01 /prosecutors-are-failing-the-victims-of-floridas-notorious-reform-school.html. Accessed April 24, 2014.

Zeitz, Joshua. *Flapper: A Madcap Story of Sex, Style, Celebrity, and the Women Who Made American Modern.* New York: Crown Publishers, 1982.

Zipf, Karin L. "In Defense of the Nation: Syphilis, North Carolina's 'Girl Problem,' and World War I." *North Carolina Historical Review* 89, no. 3 (July 2012): 276–300.

———. *Labor of Innocents: Forced Apprenticeship in North Carolina, 1715–1919.* Baton Rouge: Louisiana State Univ. Press, 2005.

INDEX